D1256022

Undergraduate Texts in Mathematics

Editors

S. Axler
F.W. Gehring
P.R. Halmos

Springer
New York
Berlin
Heidelberg
Barcelona
Budapest
Hong Kong
London
Milan
Paris
Santa Clara
Singapore
Tokyo

Undergraduate Texts in Mathematics

Anglin: Mathematics: A Concise History and Philosophy.
Readings in Mathematics.

Anglin/Lambek: The Heritage of Thales.
Readings in Mathematics.

Apostol: Introduction to Analytic Number Theory. Second edition.

Armstrong: Basic Topology.

Armstrong: Groups and Symmetry.

Axler: Linear Algebra Done Right.

Bak/Newman: Complex Analysis. Second edition.

Banchoff/Wermer: Linear Algebra Through Geometry. Second edition.

Berberian: A First Course in Real Analysis.

Brémaud: An Introduction to Probabilistic Modeling.

Bressoud: Factorization and Primality Testing.

Bressoud: Second Year Calculus.
Readings in Mathematics.

Brickman: Mathematical Introduction to Linear Programming and Game Theory.

Browder: Mathematical Analysis: An Introduction.

Buskes/van Rooij: Topological Spaces: From Distance to Neighborhood.

Cederberg: A Course in Modern Geometries.

Childs: A Concrete Introduction to Higher Algebra. Second edition.

Chung: Elementary Probability Theory with Stochastic Processes. Third edition.

Cox/Little/O'Shea: Ideals, Varieties, and Algorithms. Second edition.

Croom: Basic Concepts of Algebraic Topology.

Curtis: Linear Algebra: An Introductory Approach. Fourth edition.

Devlin: The Joy of Sets: Fundamentals of Contemporary Set Theory. Second edition.

Dixmier: General Topology.

Driver: Why Math?

Ebbinghaus/Flum/Thomas: Mathematical Logic. Second edition.

Edgar: Measure, Topology, and Fractal Geometry.

Elaydi: Introduction to Difference Equations.

Exner: An Accompaniment to Higher Mathematics.

Fine/Rosenberger: The Fundamental Theory of Algebra.

Fischer: Intermediate Real Analysis.

Flanigan/Kazdan: Calculus Two: Linear and Nonlinear Functions. Second edition.

Fleming: Functions of Several Variables. Second edition.

Foulds: Combinatorial Optimization for Undergraduates.

Foulds: Optimization Techniques: An Introduction.

Franklin: Methods of Mathematical Economics.

Hairer/Wanner: Analysis by Its History.
Readings in Mathematics.

Halmos: Finite-Dimensional Vector Spaces. Second edition.

Halmos: Naive Set Theory.

Hämmerlin/Hoffmann: Numerical Mathematics.
Readings in Mathematics.

Hilton/Holton/Pedersen: Mathematical Reflections: In a Room with Many Mirrors.

Iooss/Joseph: Elementary Stability and Bifurcation Theory. Second edition.

Isaac: The Pleasures of Probability.
Readings in Mathematics.

James: Topological and Uniform Spaces.

Jänich: Linear Algebra.

Jänich: Topology.

Kemeny/Snell: Finite Markov Chains.

Kinsey: Topology of Surfaces.

Klambauer: Aspects of Calculus.

Lang: A First Course in Calculus. Fifth edition.

Lang: Calculus of Several Variables. Third edition.

Lang: Introduction to Linear Algebra. Second edition.

Lang: Linear Algebra. Third edition.

(continued after index)

Gerard Buskes
Arnoud van Rooij

Topological Spaces
From Distance to Neighborhood

With 151 Illustrations

Springer

Gerard Buskes
Department of Mathematics
University of Mississippi
University, MS 38677
USA

Arnoud van Rooij
Department of Mathematics
Catholic University of Nijmegen
Toernooiveld
Nijmegen, 6525 ED
The Netherlands

Mathematics Subject Classification (1991): 54-01, 54A05

Library of Congress Cataloging-in-Publication Data
Buskes, Gerard.
    Topological spaces : from distance to neighborhood / Gerard
  Buskes, A. van Rooij.
      p. cm. — (Undergraduate texts in mathematics)
    Includes bibliographical references (p. -) and indexes.
    ISBN 0-387-94994-1 (alk. paper)
    1. Topological spaces. I. Rooij, A. C. M. van (Arnoud C. M.),
  1936–   . II. Title. III. Series.
  QA611.3.B87 1997
  514′.322—dc21   97-3756

Printed on acid-free paper.

Production managed by Victoria Evarretta; manufacturing supervised by Jacqui Ashri.
Photocomposed copy prepared from the authors' LaTeX files.
Printed and bound by Maple-Vail Book Manufacturing Group, York, PA.
Printed in the United States of America.

9 8 7 6 5 4 3 2 1

ISBN 0-387-94994-1 Springer-Verlag New York Berlin Heidelberg   SPIN 10557376

Preface

This book is a text, not a reference, on Point-set Topology. It addresses itself to the student who is proficient in Calculus and has some experience with mathematical rigor, acquired, e.g., via a course in Advanced Calculus or Linear Algebra.

To most beginners, Topology offers a double challenge. In addition to the strangeness of concepts and techniques presented by any new subject, there is an abrupt rise of the level of abstraction. It is a bad idea to teach a student two things at the same moment. To mitigate the culture shock, we move from the special to the general, dividing the book into three parts:

1. The Line and the Plane
2. Metric Spaces
3. Topological Spaces.

In this way, the student has ample time to get acquainted with new ideas while still on familiar territory. Only after that, the transition to a more abstract point of view takes place.

Elementary Topology preeminently is a subject with an extensive array of technical terms indicating properties of topological spaces. In the main body of the text, we have purposely restricted our mathematical vocabulary as much as is reasonably possible. Such an enterprise is risky. Doubtlessly, many readers will find us too thrifty. To meet them halfway, in Chapter 18 we briefly introduce and discuss a number of topological properties, but even there we do not touch on paracompactness, complete normality, and extremal disconnectedness—just to mention three terms that are not really esoteric.

In a highly abstract topic like ours, it aids a student to focus on a central theme. The theme of our book is convergence. We show how, for $\mathbb{R}^n$ and for metric spaces in general, concepts such as "continuous" and "closed" can be described in terms of convergent sequences. After that, in any given set X we introduce convergence of nets relative to any given collection ω of subsets of X. This convergence leads in a natural way to the notion of a topology. The idea behind this somewhat unconventional approach is threefold.

First, it shows that the definition of "topology" is less artificial than it seems to be. Without this preparation, the definition appears to stem from an arbitrary selection of properties of the system of open sets in $\mathbb{R}^n$, and it is not clear why precisely *these* properties are the relevant ones. (The reader who finds this a digression can skip Chapter 11; in Chapter 12, the definition and some basic facts are repeated without the motivation.)

Second, it relegates the notion of "topology" to a place in the second rank. When one studies a topological space, often the topology itself is less relevant than a subbase for it (the collection ω in the situation described above). A case in point is the product topology on a Cartesian product of topological spaces: all that really matters is a subbase, and the fact that this subbase generates a topology is quite immaterial.

Third, convergent nets form a very useful tool in Topology, deserving much more attention than they generally get.

We do not assume previous knowledge of the axiomatic approach. As, however, a rigorous theory of topological spaces must have a firm base in Analysis, we start with a brief axiomatic treatment of the real-number system, explaining what axioms are and what purpose they serve.

We do not assume previous knowledge of Set Theory either. (Indeed, to be on the safe side, we have added a chapter on countability.) On the other hand, Topology unavoidably leads to nontrivial set-theoretic problems. Accordingly, in connection with the Tychonoff Theorem, we pay close attention to the Axiom of Choice and Zorn's Lemma and their role in mathematics.

The pace of this book is relaxed with a gradual acceleration. For instance, the first three chapters and part of Chapter 4 can be relegated to home reading for a well-prepared student. However, the easy initial pace makes the first nine chapters a balanced course in metric spaces for undergraduates. The book contains more than enough material for a two-semester graduate course.

As with all mathematical learning, a substantial amount of practice is indispensable. We offer exercises of varying degrees of difficulty. Some are routine, others illustrate results of the text, and yet others go beyond the text. We have carefully crafted these exercises. Accordingly, one will find many of them, in particular the complicated ones, sectioned into more digestible pieces with hints.

Finally, in most chapters we present an "extra," a brief foray outside Topology. A beginning student is apt to consider each branch of mathematics as an autonomous unit, isolated from the rest, and also to think that mathematics is a museum piece, something created in olden times by our forefathers, that can be seen and even studied, but not touched. Our purpose of the extras is to illustrate the many connections between Topology and other subjects, such as Analysis and Set Theory. Also, in our extras we try to show that Topology was and still is built by individuals, who sometimes made mistakes. We encourage the reader to consider these extras to be part of the course. The extras are extra, not extraneous.

Contents

PART

The Line and the Plane

CHAPTER

What Topology Is About

Topological Equivalence

1.1 Question

Which picture does not go with the others?

1.2

The last one, of course, but it is not so easy to describe the common feature of the first six shapes that is lacking in the seventh. A satisfactory description can be given in the language of *Topology*, the subject matter of this book.

To give you some idea of what topology is, we return for a moment to plane geometry. Suppose you have drawn a triangle with sides of 13, 14, and 15 inches and by measuring you have found that it has an angle of 54°. Then you know that *every* triangle with sides of 13, 14, and 15 inches must have a 54° angle, because all such triangles are congruent. Their positions and orientations in the plane do not matter to a geometer, as they would to a surveyor. Using an arbitrarily chosen term, we will say that the geometer's point of view is "higher" than the surveyor's. The

3

surveyor distinguishes among the following triangles, the geometer does not:

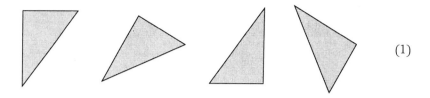

(1)

One can adopt a higher point of view than the geometer's: For certain purposes, there is no sense in distinguishing

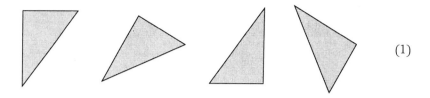

(2)

or (still higher)

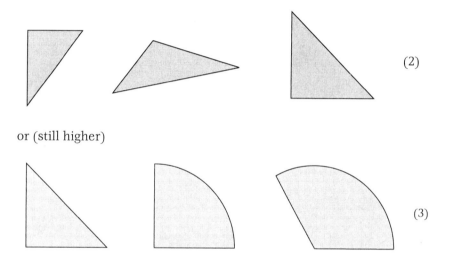

(3)

At this stage, the shapes differ considerably, but if you draw them on pieces of rubber instead of paper, you can obtain them from each other by stretching and bending:

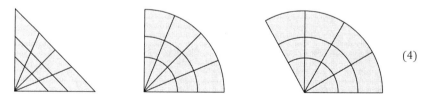

(4)

Here, we arrive at the heart of the matter. Topology is the branch of mathematics in which the differences between the shapes of (3) are irrelevant, precisely as those between the shapes of (1) are irrelevant in geometry. The topologist's point of view is "higher" than the geometer's.

Carrying the deformations a bit farther, we obtain shapes that no longer have anything triangular about them:

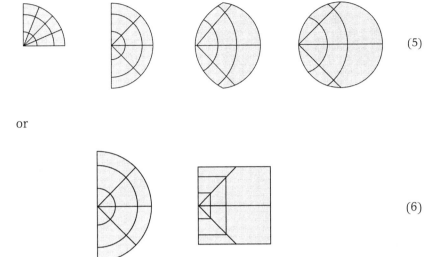

(5)

or

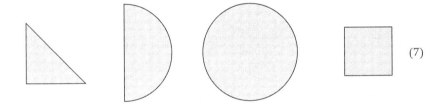

(6)

Thus, in the eyes of the topologist, the shapes

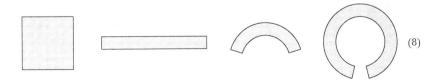

(7)

are the same; let us call them topologically equivalent [as the triangles in (1) may be called geometrically equivalent].

From the square, by stretching and bending we obtain another series of shapes:

(8)

But no amount of stretching will produce a closed ring:

Here, something else is required, such as gluing two edges together. The
ring and the square are not topologically equivalent; they are as different to the topologist as the square and the triangle are to the geometer.
Similarly, the square may be stretched

but to obtain two triangles

one would have to tear the rubber: The pair of triangles is not topologically
equivalent to the square.

1.3

What has all this to do with mathematics? Stretching a piece of rubber is
hardly a mathematical operation. But the grids sketched in (4), (5), and
(6) suggest how the concept of topological equivalence can be defined
mathematically. Compare the first shape of (4) and the last of (5):

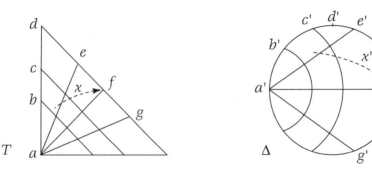

Deforming the triangle T into the disk Δ yields a one-to-one correspondence between the points $a, b, c, \ldots$ of the first and the points $a', b', c', \ldots$ of the second, respectively. In other words, we have a bijection $\varphi : T \to \Delta$. Due to the fact that our deformations do not allow tearing, if a point x of T approaches the point f, then the corresponding point x' of Δ comes close to f'. Mathematically speaking, this means that the bijection φ is continuous. Similarly, its inverse, $\varphi^{-1} : \Delta \to T$, is continuous.

1.4
We define (provisionally): Two subsets of the plane, A and B, are *topologically equivalent* if there exists a bijective map $\varphi : A \to B$ such that both φ and its inverse are continuous. The first six shapes drawn in 1.1 are topologically equivalent to each other but not to the seventh. One can give an exact proof of that once one has exact descriptions of the shapes themselves.

1.5
The same definition is meaningful for subsets of three-dimensional space or of a line. For instance, the intervals $(0, 1)$ and $(-\infty, 0)$ of $\mathbb{R}$ are topologically equivalent: The logarithm yields a continuous bijection $(0, 1) \to (-\infty, 0)$ with a continuous inverse. Somewhat harder to see is that the intervals $[0, 1]$ and $(0, 1)$ are *not* topologically equivalent. (Suppose they are. Take a bijection $\varphi : [0, 1] \to (0, 1)$ such that both φ and φ^{-1} are continuous. On $[0, 1]$, every continuous function attains a smallest value. Let a be the smallest value of φ. All values of φ lie in $(0, 1)$, so $a \in (0, 1)$. It follows that $\frac{1}{2} a \in (0, 1)$, so that $\frac{1}{2} a$ must be a value of φ. But then, a cannot be the *smallest* value of φ.)

Actually, we will go much farther and develop a theory that has the plane, three-dimensional space, and the real line as special cases. First, we must reconsider the concept of continuity.

Continuity and Convergence

1.6
From calculus you are, of course, familiar with continuity. However, most calculus texts treat continuity in a very restricted way. Continuity of a function of one variable is usually defined only if the domain of the function is an interval. A legitimate question in Topology is whether the sets $\mathbb{Q} \cap (0, 1)$ and $\mathbb{Q} \cap [0, 1]$ are topologically equivalent. Contrary to what you would expect after reading 1.5, they are, but a proof requires knowledge of continuous functions with domain $\mathbb{Q} \cap (0, 1)$. Here, calculus books let us

down. And they provide hardly any information at all concerning maps between subsets of $\mathbb{R}^2$.

Therefore, we start with a study of continuity on arbitrary subsets of $\mathbb{R}$; in Chapter 3, we look at $\mathbb{R}^2$. You may already know all this from an Advanced Calculus course. In that case you may just skim over our first three chapters. (Have a look at Theorem 1.12. If that is familiar terrain, you may as well skip the balance of Chapter 1 except for 1.14, where we fix some notations.)

1.7

Let D be a subset of $\mathbb{R}$ and let f be a function on D.

Take a point a of D. We say that f is *continuous at a* if

$$\left.\begin{array}{l} \text{for every positive number } \varepsilon \\[4pt] \text{there exists a positive } \delta \text{ such that} \\[4pt] \text{for all } x \text{ in } D \cap (a-\delta, a+\delta) : \ |f(x)-f(a)| < \varepsilon. \end{array}\right] \qquad (*)$$

f is called *continuous* (without mention of any specific point of D) if f is continuous at every point of D.

Examples 1.8

These definitions seem humdrum, but look at the following examples.

(i) Define a function f whose domain is $\mathbb{Q}$ by

$$f(x) := 1 \quad \text{if} \quad x \in \mathbb{Q}, \ x < \sqrt{2},$$
$$f(x) := 2 \quad \text{if} \quad x \in \mathbb{Q}, \ x > \sqrt{2}.$$

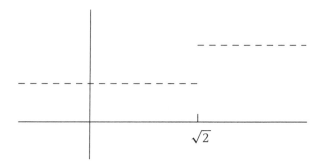

$\sqrt{2}$ is not a rational number, so the above formulas determine $f(x)$ unambiguously for every element x of $\mathbb{Q}$.

Let $a \in \mathbb{Q}$. Then $a \neq \sqrt{2}$, so the number $\delta_0 := |\sqrt{2}-a|$ is positive. The interval $(a-\delta_0, a+\delta_0)$ lies either completely to the left or completely to the right of $\sqrt{2}$. In either case, f is constant on $\mathbb{Q} \cap (a-\delta_0, a+\delta_0)$:

$$\text{for all } x \text{ in } \mathbb{Q} \cap (a-\delta_0, a+\delta_0) : \quad f(x) = f(a).$$

Hence, no matter what positive number ε one takes, the last line of (∗) in 1.7 will be satisfied by $\delta = \delta_0$. Consequently, f is continuous at a.

But a was an arbitrary element of the domain of f. Thus, f is a continuous function.

(ii) Let g be any function whose domain is $\mathbb{N}$; we show that it is necessarily continuous at the point 3 of $\mathbb{N}$. Indeed, let $\varepsilon > 0$ be given; we take $\delta = \frac{1}{2}$. For any x in $\mathbb{N} \cap (3-\delta, 3+\delta)$, we have $x = 3$, hence $g(x) = g(3)$, hence $|g(x)-g(3)| < \varepsilon$.

In the same way, of course, *every* function with domain $\mathbb{N}$ is continuous at *every* point of $\mathbb{N}$.

1.9

Closely related to the notion of continuity is the one of convergence of a sequence. We recall a few definitions and elementary facts.

A *sequence* is a function whose domain is $\mathbb{N}$. The sequence whose value at n is x_n will often be denoted by

$$x_1, x_2, x_3, \ldots$$

or

$$(x_n)_{n\in\mathbb{N}}.$$

For $N \in \mathbb{N}$, the N-th *tail* of the sequence $x_1, x_2, \ldots$ is the sequence

$$x_N, x_{N+1}, x_{N+2}, \ldots.$$

Let $(x_n)_{n\in\mathbb{N}}$ be a sequence, and let a be a number. We say that the sequence *converges to* a, or that a is a *limit* of the sequence if

$$\left.\begin{array}{l} \text{for every } \varepsilon > 0, \text{ the interval } (a-\varepsilon, a+\varepsilon) \\[6pt] \text{contains a tail of the sequence.} \end{array}\right] \tag{1}$$

This is the case if and only if

$$\left.\begin{array}{l} \text{for every } \varepsilon > 0, \\[4pt] \text{there exists an } N \text{ in } \mathbb{N} \text{ such that} \\[4pt] |x_n-a| < \varepsilon \text{ as soon as } n \geq N. \end{array}\right] \tag{2}$$

The latter formulation may be closer to the one you are used to, but both definitions mean precisely the same thing.

Statements (1) and (2) will mostly be abbreviated as

$$x_n \to a.$$

1.10

As you know from Calculus, *a sequence has at most one limit.* Thus, we may speak of "the" limit of a converging sequence $(x_n)_{n \in \mathbb{N}}$ and denote it by

$$\lim_{n \to \infty} x_n.$$

There are a few simple rules for handling converging sequences:

Theorem 1.11

Let $a, x_1, x_2, \ldots, b, y_1, y_2, \ldots \in \mathbb{R}$.

  (i) *If $x_n \to a$, then $|x_n - a| \to 0$, and vice versa.*
 (ii) *If $x_n \to a$ and $|y_n - b| \leq |x_n - a|$ for all n, then $y_n \to b$.*
(iii) *If $x_n \to a$ and $y_n \to b$, then $x_n + y_n \to a + b$ and $x_n y_n \to ab$.*
 (iv) *If $x_n \to a$ and if $x_n \geq 0$ for all n, then $a \geq 0$.*
  (v) *If $x_n \to a$, $y_n \to b$ and $x_n \geq y_n$ for all n, then $a \geq b$.*

The following theorem establishes a connection between continuity and convergence that will be a recurring theme in this book.

Theorem 1.12

Let $D \subset \mathbb{R}$, let f be a function whose domain is D, and let $a \in D$. Then the conditions (α) and (β) are equivalent.

(α) *f is continuous at a.*
(β) *For every sequence $(x_n)_{n \in \mathbb{N}}$ in D with $x_n \to a$, we have $f(x_n) \to f(a)$.*

Proof

$(\alpha) \implies (\beta)$ is easy: Let $x_1, x_2, \ldots \in D$, $x_n \to a$. We wish to prove $f(x_n) \to f(a)$. Thus, let $\varepsilon > 0$. There exists a $\delta > 0$ such that

$$\text{for all } x \text{ in } D \cap (a - \delta, a + \delta) : \quad |f(x) - f(a)| < \varepsilon.$$

As $x_n \to a$ there exists an $N \in \mathbb{N}$ with

$$\text{for all } n \geq N : \quad |x_n - a| < \delta.$$

For $n \geq N$, we then have $x_n \in D \cap (a - \delta, a + \delta)$ and, consequently, $|f(x_n) - f(a)| < \varepsilon$. This proves (β).

$(\beta) \implies (\alpha)$ is harder. Assume (β). Let $\varepsilon_0 > 0$. We need a positive number δ such that

$$\text{for all } x \text{ in } D \cap (a - \delta, a + \delta) : \quad |f(x) - f(a)| < \varepsilon_0. \tag{$*$}$$

Consider the following hypothesis:

$$\text{No positive number } \delta \text{ satisfies } (*). \tag{H}$$

We are done if we can show that this hypothesis is untenable, so, for a while, let us *suppose* it is true.

Then the positive number $\delta = \frac{1}{100}$ does not satisfy (∗), i.e., it is not true that for every x in $D \cap (a - \frac{1}{100}, a + \frac{1}{100})$ we have $|f(x) - f(a)| < \varepsilon_0$. Putting it differently, there must exist an x in $D \cap (a - \frac{1}{100}, a + \frac{1}{100})$ without $|f(x) - f(a)| < \varepsilon_0$; or there must exist an x in $D \cap (a - \frac{1}{100}, a + \frac{1}{100})$ with $|f(x) - f(a)| \geq \varepsilon_0$.

The choice $\delta = \frac{1}{100}$ we made above was quite arbitrary. For every $n \in \mathbb{N}$, we can observe that $\delta = \frac{1}{n}$ does not satisfy (∗), so that there must exist an x_n in $D \cap (a - \frac{1}{n}, a + \frac{1}{n})$ with $|f(x_n) - f(a)| \geq \varepsilon_0$.

We have now obtained a sequence $x_1, x_2, \ldots$ in D. This sequence converges to a since $|x_n - a| < \frac{1}{n}$ for all n. But the sequence $f(x_1), f(x_2), \ldots$ cannot converge to $f(a)$ since $|f(x_n) - f(a)| \geq \varepsilon_0$ for all n. We see that (β) of the theorem cannot hold.

We have obtained a *contradiction* from the hypothesis (H). Thus, we have refuted (H) and our proof is finished. ∎

As the function $x \mapsto |x|$ is continuous, by the implication $(\alpha) \implies (\beta)$ we have

$$\text{if } x_n \to a, \text{ then } |x_n| \to |a|.$$

Similarly,

$$\text{if } x_n \to a, \text{ then } e^{x_n} \to e^a,$$

and so on.

1.13
By way of an application we show the product of continuous functions to be continuous. Let f and g be functions on a domain $D \subset \mathbb{R}$ and let both be continuous at a point a of D. Define $h : D \to \mathbb{R}$ by $h(x) = f(x)g(x)$ ($x \in D$). Then h is continuous at a. Indeed, for every sequence $(x_n)_{n \in \mathbb{N}}$ in D that converges to a, we have $f(x_n) \to f(a)$ and $g(x_n) \to g(a)$ [$(\alpha) \implies (\beta)$ of 1.12], and, therefore, $h(x_n) \to h(a)$ [1.11(iii)]. Then h is continuous at a according to $(\beta) \implies (\alpha)$ of 1.12.

A Few Conventions

1.14
We close the first chapter with a list of notations and terminology to be used in the sequel.

(i) The symbol

$$:=$$

will indicate a definition. For instance,

$$X := \{x \in \mathbb{R} : x^3 + x^2 < 6\}$$

means: We define X to be the set $\{x \in \mathbb{R} : x^3 + x^2 < 6\}$.
  (ii) If A and B are subsets of X, then

$$A \backslash B := \{x \in X : x \in A \text{ but } x \notin B\}.$$

 (iii) We will often deal with maps between sets. Whenever we write "f is a map of X into Y" or

$$f : X \to Y$$

we mean that X and Y are sets and that f is a map whose domain is X and whose values lie in Y. In particular, $f(x)$ makes sense for *every* x in X (but we do not require that every element of Y be a value of f). Thus, the sine function is a map $\mathbb{R} \to \mathbb{R}$; the logarithm is not.
   A *function* is a map whose values are real numbers; its domain may be any set.
 (iv) If A is a subset of a set X, the *indicator* of A is the function $\mathbf{1}_A : X \to \mathbb{R}$ defined by

$$\mathbf{1}_A(x) := 1 \quad \text{if} \quad x \in A,$$

$$\mathbf{1}_A(x) := 0 \quad \text{if} \quad x \in X, \ x \notin A.$$

Occasionally, we denote by $\mathbf{1}$ the constant function whose value is 1.
  (v) Suppose we have a map $f : X \to Y$ and a subset A of X. The *restriction of f to A* is the map $f|_A$ of A into Y:

$$(f|_A)(x) := f(x) \quad \text{if} \quad x \in A.$$

f and $f|_A$ are distinct functions (if $A \neq X$) and should not be confused. For example, $\sin|_{(0,\pi/2)}$ is increasing, but $\sin$ is not! Observe: If $A \subset X \subset \mathbb{R}$ and if f is a continuous function on X, then $f|_A$ is also continuous.
 (vi) The formula

$$x \longmapsto x^2 + 6x - 1 \qquad (0 \leq x \leq 1) \tag{1}$$

indicates the function $f : [0, 1] \to \mathbb{R}$ defined by

$$f(x) := x^2 + 6x - 1 \qquad (0 \leq x \leq 1).$$

In a formula such as (1), we will always mention the domain; it is bad manners to talk of "the function $x \longmapsto x^2 + 6x - 1$" and leave it to the reader to figure out what the domain should be.
(vii) If $f : X \to \mathbb{R}$ and $g : Y \to \mathbb{R}$ are functions, then $f + g$ and fg are the functions

$$x \longmapsto f(x) + g(x) \qquad (x \in X \cap Y)$$

and

$$x \longmapsto f(x)g(x) \qquad (x \in X \cap Y),$$

respectively.

(viii) For maps $f : X \to Y$ and $g : Y \to Z$, we have a *composite map* $g \circ f : X \to Z$:

$$(g \circ f)(x) = g(f(x)) \qquad (x \in X).$$

Observe: if $f : X \to Y$, $g : Y \to Z$, and $h : Z \to U$, then the maps $(h \circ g) \circ f$ and $h \circ (g \circ f)$ are the same; we will denote this map by

$$h \circ g \circ f.$$

(ix) If $f : X \to Y$ and $A \subset X$, then

$$f(A) := \{f(x) : x \in A\};$$

this $f(A)$ is a subset of Y.

(x) In 1.9 we have defined a sequence to be a function on $\mathbb{N}$. We extend this definition. For any set X, a *sequence* of elements of X is a map of $\mathbb{N}$ into X. The sequence $n \mapsto x_n$ is also denoted

$$x_1, x_2, \ldots \quad \text{or} \quad (x_n)_{n \in \mathbb{N}}.$$

If X is a set, a *family* of elements of X is a map of some set into X. A family $s \mapsto x_s$ ($x \in S$) is often indicated as

$$(x_s)_{s \in S}.$$

(xi) If A and B are sets, by the *Cartesian product*

$$A \times B$$

we mean the set of all pairs (a, b) with $a \in A$, $b \in B$:

$$A \times B := \{(a, b) : a \in A, b \in B\}.$$

(xii) Special attention is asked for the *implication arrow*

$$\implies.$$

It is used as a conjunction between two statements or formulas. By

$$\mathcal{A} \implies \mathcal{B}$$

(where $\mathcal{A}$ and $\mathcal{B}$ are entire sentences) we mean "$\mathcal{B}$ is a logical consequence of $\mathcal{A}$," or "whenever we have $\mathcal{A}$, $\mathcal{B}$ necessarily follows." This does not say anything about the validity of $\mathcal{A}$ and $\mathcal{B}$ themselves. For instance, if x is a real number, then

$$x > 100 \implies x > 5$$

regardless of whether x is, indeed, larger than 100.

This may sound obvious, but notice that our use of $\implies$ implies the truth of strange formulas such as

$$3 > 100 \implies 3 > 5$$

and also

$$8 > 100 \implies 8 > 5$$

Using the implication arrow, a phrase like

$$|x_n - a| < \varepsilon \text{ as soon as } n > N$$

(occurring in the definition of convergence) may be abbreviated as

$$n > N \implies |x_n - a| < \varepsilon.$$

We also have a double arrow: The formula

$$\mathcal{A} \iff \mathcal{B}$$

means: $\mathcal{A} \implies \mathcal{B}$ and $\mathcal{B} \implies \mathcal{A}$. Example: If $x \in \mathbb{R}$, then

$$x^3 = 8 \iff x = 2.$$

Extra: Topological Diversions

Many puzzles and games have topological connotations.
   A famous brainteaser is the "handcuff puzzle":

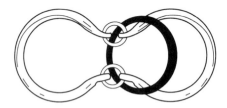

Can you take off the black ring?
   Topologically, there is no problem. The black ring is not linked to the rest of the puzzle; it would come off if you could stretch it. But the puzzle is made of metal. It does not stretch, and we are dealing with geometry, not topology. Geometrically, there *is* a problem.
   A more complicated puzzle of the same type is this one:

It was invented in China, according to the story, by a famous warrior, Hung Ming (181-234), who wanted to keep his wife entertained during his absences. It appeared in Europe in the seventeenth century and got the name "Meleda."

The following notorious puzzle is purely topological. Imagine three houses and three wells. Can you connect each house with each well without having two connections crossing?

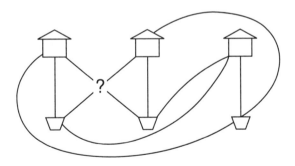

In the sketch, we have indicated eight of the nine required connections, but you can see that it is now impossible to connect the third house with the third well without having two lines cross.

This problem is topological in that the positions of the houses and the wells are irrelevant. If you can solve it for the configuration below

then, by drawing your solution on a rubber sheet and deforming it, you see that you can solve it for the originally given situation.

Using topological arguments, one can prove that there is *no* solution, when you draw your picture in a plane or on the surface of a ball. But if the houses and the wells are on a planet that has the shape of a donut,

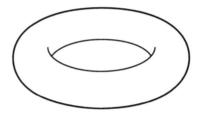

there is a way around it!

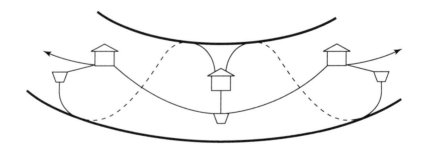

Actually, now you can add a fourth house and connect it (without crossovers) to all three wells. Try it!

The "problem of the Koenigsberg bridges" was raised by the great Swiss mathematician Leonhard Euler in 1736. The city of Koenigsberg (now Kaliningrad) where he lived at the time was built on both sides of a river and on two islands. The four parts were connected by seven bridges:

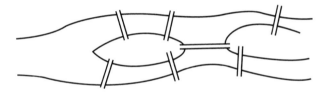

Question: Can you make a walk that crosses every bridge once but not more than one?

Again, a topological problem. The geographical details are of no importance. Without any loss, you may stylize the picture like

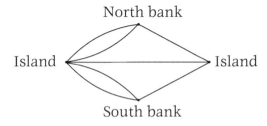

The entertaining game "sprouts" is described by Martin Gardner[1] in *Mathematical Carnival* (Vintage Books, 1977).

The game begins with a number of spots on a sheet of paper. The players take turns making moves. A move consists of drawing a line beginning at a spot and ending at a spot (possibly the same) and placing a new spot anywhere along the line. The line may be straight or curved, but is not allowed to cross itself or any previously drawn line or pass through an already existing spot. No spot may have more than three lines emanating from it. The first player who cannot make a move loses.

Starting with two dots, a game could develop as follows

The game cannot go on indefinitely. If there are n spots at the start, there can be at most $3n-1$ moves. (Can you prove that?)

---

[1] In case you do not know of Martin Gardner, run to the library. You will need him next time some barbarian tells you that mathematics can interest only mathematicians.

Further Reading

Stewart, I., The Topological Dressmaker, *Scientific American*, July 1993, 110-112.

Exercises

1.A. Recall that for a topologist the shapes sketched in (1)-(8) are essentially identical; because they can be obtained from one another by continuous deformations, they are "topologically equivalent." As J.L. Kelley expresses it in his famous text-book *General Topology*: "A topologist is a man who doesn't know the difference between a doughnut and a coffee cup." (The book was published in 1955; the spelling is outdated and so is the notion that there are no women topologists.)

Similarly, a topologist would not distinguish between a straight line, such as a letter " l ," and a curved one, like an " S ." Show that in the eyes of a topologist

$$ONE$$

is the same as

$$TWO$$

1.B. Imagine you take a strip of paper

and stick the ends together.

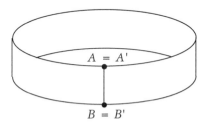

What you get is a ring. If you cut it lengthwise,

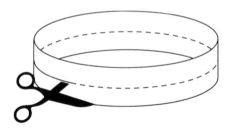

you obtain two narrow rings.

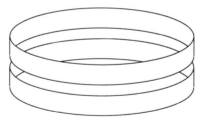

The following variation is harder to visualize. Take the strip, give it a twist,

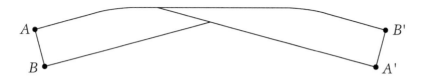

and then stick the ends together so that A covers B' and B covers A':

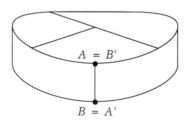

This time you get a so-called "Moebius strip." What happens if you cut it lengthwise?

Think about it first. Then take a paper strip, some Scotch tape, and a pair of scissors, and carry out the construction.

1.C. A rubber balloon

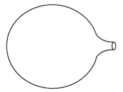

can be turned inside out. An ordinary balloon looks more or less like a ball, but imagine a balloon shaped like a tire or (the surface of) a donut:

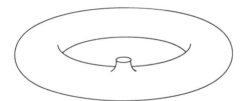

Can you turn it inside out? The material is supposed to be extremely flexible and you may widen the hole as much as you like, but without tearing anything.

Again, first think it over, then try it. For an actual model, rubber is too stiff. You can make a serviceable tire balloon from a square piece of clear plastic, 10 inches by 10 inches, say. You fold it and tape it together as follows:

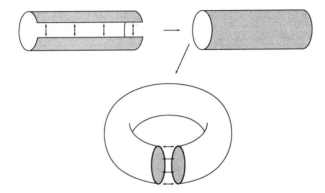

(As befits true topologists, our sketches are not to scale.) You have made a very flat tire. Make a hole in it, about 2 inches across, and strengthen the hole with tape:

Now see if it is possible to completely evert the tire through the hole.

1.D. This exercise is a sequel to the preceding one. Unless you have finished that one you are not allowed to read on.

You will have discovered that the tire can be everted. The process gives you a crumpled wad of plastic, but after some smoothing, you can recognize it as being a tire, just like the first. No surprise here: If you turn a sock inside out you get a sock, not a glove. Still, something strange is occurring.

On a tire there are two types of circles: meridians and parallels.

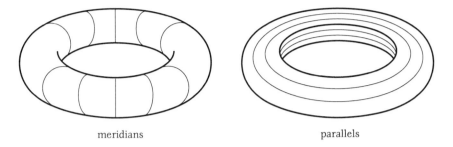

meridians   parallels

Take the tire you have made and paint a few meridians on it. Evert the tire. The painted circles are on the inside, but you can still see them because the plastic is clear. (If you have followed the instructions!) What has happened?

1.E. We return for a moment to the Moebius strip. An ordinary piece of paper has two sides. Two ants on different sides can get together only if one of them crawls over the edge. But on a Moebius strip they can find each other without such acrobatics, as illustrated in this famous lithograph "Moebius Band II" by M.C. Escher (©1996 Cordon Art-Baarn-Holland. All rights reserved.)

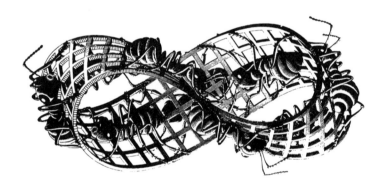

There is a limerick about the Moebius strip that is well known in mathematical circles and has to do with Exercise 1.B:

> A mathematician confided
> That a Moebius strip is one-sided
> And you get quite a laugh
> When you cut it in half:
> It stays in one piece when divided.

The remarkable properties of the Moebius strip have inspired more poetry, such as

> The topologist's child was quite hyper
> Till she wore a Moebius diaper.
> The mess on the inside
> Was thus on the outside:
> It was easy for someone to wipe her.

The exercise is this: Show that the problem of the three houses and the three wells on the Moebius strip can be solved. (It cannot on the ring made at the beginning of 1.B !)

It is understood that the lines connecting the houses and the wells are not to be drawn *on* the surface, any more than, say, the line $y = x$ lies on top of the $x - y$ plane.

2

CHAPTER

Axioms for $\mathbb{R}$

As you have seen in the previous chapter, continuity and convergence are basic for topology, and calculus is not sufficient as a background. The latter statement has a double meaning. We need more calculus-like theorems, especially on functions of several variables, but that is not all: Topology also requires a more precise kind of reasoning than an introductory Calculus course. We will have to prove all of our theory and we cannot afford to rely on pictures (although they will be an invaluable aid). In this chapter, we lay the foundations for a rigorous theory in the form of a system of axioms.

From Advanced Calculus or an equivalent course, you may already have experience with the axiomatic method. In that case, it will hardly be necessary for you to spend much energy on this chapter. You may still want to look it over; quite possibly our terminology will differ from what you have seen before and there is a good chance that the "Postscript" contains something that is new to you.

2.1

Contrary to what you may expect, we are not going to define the term "real number." You may view real numbers as points of a line or as endless strings of decimals or in any other way you like. Actually, what real numbers are will be irrelevant to us, just as the shapes of the pieces are irrelevant to a chess player.

What does matter to him are the rules of the game. Like him, we will stick to a list of rules. Our rules will be the "Axioms of Analysis," certain mathematical statements in which the term "real number" occurs. We hope you will find that with your personal interpretation of the term, these statements are true. We then proceed to draw conclusions from these axioms. While doing so we use only strict logic and not, for instance, our intuitive idea of real numbers. Thus, we will be certain that, having agreed with the axioms, you will agree with the conclusions, even though your "real numbers" may be quite different from ours.

  If you dislike the rules of chess, you may adopt other ones. There is nothing criminal about that, but you will not be playing chess. Similarly, if you do not like our axioms, nothing prohibits you from inventing other ones and developing a new branch of mathematics.

2.2

Axiom I.
ℝ is a set. Its elements are called *numbers*. 0 and 1 are numbers, $0 \neq 1$. To any pair x, y of numbers there are assigned a number $x+y$, called their *sum*, and a number xy, their *product*, subject to the following rules.

  (a) If $x, y \in \mathbb{R}$, then $x+y = y+x$.
  (b) If $x, y, z \in \mathbb{R}$, then $(x+y)+z = x+(y+z)$.
  (c) If $x \in \mathbb{R}$, then $x+0 = x$.
  (d) If $x \in \mathbb{R}$, then there exists a unique element $-x$ of ℝ such that
      $x+(-x) = 0$.
  (e) If $x, y \in \mathbb{R}$, then $xy = yx$.
  (f) If $x, y, z \in \mathbb{R}$, then $(xy)z = x(yz)$.
  (g) If $x, y, z \in \mathbb{R}$, then $x(y+z) = (xy) + (xz)$.
  (h) If $x \in \mathbb{R}$, then $x1 = x$.
  (i) If $x \in \mathbb{R}$ and $x \neq 0$, then there exists a unique element x^{-1} in ℝ
      such that $xx^{-1} = 1$.

Our list of axioms is, as yet, far from complete, but let us sit back for a moment and reflect on what we have.

  First, do you find the statements (a)-(i) reasonable? If not, then you are in trouble because the entire content of this book is built upon them.

  But let us assume you are willing to accept them. You do not really need absolute faith in their validity; it suffices for you to go along with us while we explore their consequences.

  Starting from Axiom I, one can begin to develop a rudimentary theory of real numbers. One introduces notations and abbreviations such as

$$2 := 1+1,$$

$$x^2 := xx,$$

$$x-y := x + (-y),$$

$$x+y+z := x + (y+z)$$

and proves formulas like

$$(x+y)^2 = x^2 + 2xy + y^2,$$
$$x^2 - y^2 = (x+y)(x-y),$$
$$xy = 0 \quad \Longrightarrow \quad x = 0 \quad \text{or} \quad y = 0.$$

As an illustration, we prove that $x0 = 0$ for every x:

$$x0 \overset{(c)}{=} (x0) + 0 \overset{(d)}{=} (x0) + (x+(-x))$$

$$\overset{(b)}{=} (x0+x) + (-x) \overset{(h)}{=} (x0+x1) + (-x)$$

$$\overset{(g)}{=} (x(0+1)) + (-x) \overset{(a)}{=} (x(1+0)) + (-x)$$

$$\overset{(c)}{=} (x1) + (-x) \overset{(h)}{=} x + (-x) \overset{(d)}{=} 0.$$

You might object that a formula like

$$x0 = 0 \qquad\qquad\qquad\qquad\qquad (*)$$

is too obvious to require a proof, and you might well be right. The purpose of the above maneuver was not to establish the validity of the formula but to show that it is a necessary consequence of Axiom I.

2.3
Many formulas and theorems can be derived from the axiom, but they are all of the same elementary nature as $(*)$, above. It would be a terrible waste of time to actually carry out the proofs. Instead, let us consider the axiom ruling the ordering $\leq$:

Axiom II.
$\leq$ is a binary relation in $\mathbb{R}$ satisfying:

(a) If $x \leq y$ and $y \leq z$, then $x \leq z$.
(b) If $x \leq y$ and $y \leq x$, then $x = y$.
(c) If $x \in \mathbb{R}$, then $x \leq x$.
(d) If $x, y \in \mathbb{R}$, then $x \leq y$ or $y \leq x$.
(e) If $x \leq y$ and $z \in \mathbb{R}$, then $x+z \leq y+z$.
(f) If $x \leq y$ and $0 \leq u$, then $xu \leq yu$.

At this stage, again one puts forward some notations:

$$x < y \quad \Longleftrightarrow \quad x \leq y \quad \text{and} \quad x \neq y,$$
$$[a, b] := \{x : a \leq x \quad \text{and} \quad x \leq b\},$$

$$\begin{cases} |x| := x & \text{if } 0 \le x, \\ |x| := -x & \text{if } x < 0, \end{cases}$$

and one proves, e.g.,

if $x < y$ and $y \le z$, then $x < z$,

if $x < y$, then $-y < -x$,

$0 < 1$,

$|x+y| \le |x| + |y|$ for all x, y in ℝ.

2.4
We use this opportunity to introduce a notation that may be new to you. For any $x, y \in ℝ$, by

$$x \vee y$$

we indicate the larger of the numbers x and y, by

$$x \wedge y$$

the smaller. Thus, $0 \vee 1 = 1$ and $0 \wedge 1 = 0$. For every $x \in ℝ$, we have $x \vee (-x) = |x|$ and $x \wedge (-x) = -|x|$.

$\vee$ and $\wedge$ are operations in ℝ, just like addition and multiplication. Actually, they obey the commutativity and associativity laws: for all $x, y, z \in ℝ$, we have

$$x \vee y = y \vee x, \quad (x \vee y) \vee z = x \vee (y \vee z),$$

$$x \wedge y = y \wedge x, \quad (x \wedge y) \wedge z = x \wedge (y \wedge z).$$

Accordingly, we will write "$x \vee y \vee z$" and "$x \wedge y \wedge z$" without provoking misunderstanding.

2.5
Once more, many conclusions can be drawn from the axioms, but very few are interesting. The upshot of it all is that most identities and inequalities you know from high school algebra are consequences of Axioms I and II.

"Most," not "all." A formula such as

$$\sqrt{2}\,\sqrt{3} = \sqrt{6}$$

does *not* follow from our axioms, simply because they are not powerful enough to ensure the existence of $\sqrt{2}$, i.e., of a solution to the equation $x^2 = 2$. Also, the axioms we have are insufficient to serve as a base for calculus. Nothing would prevent us at this stage from defining derivatives and integrals, but we would be unable to prove the existence of, say, $\int_1^2 x^{-1} dx$.

As a first step to overcome these difficulties, we present an axiom establishing the natural numbers; before that, a definition. We call a subset X of $\mathbb{R}$ *hereditary* if it has the property

$$\text{if } x \in X, \text{ then } x+1 \in X.$$

Examples of hereditary sets are $(0, \infty)$, $\{1\} \cup [2, \infty)$, and $\mathbb{R}$ itself.

Axiom III.
$\mathbb{N}$ is a subset of $\mathbb{R}$ with the following properties:

(a) $1 \in \mathbb{N}$.
(b) If $x, y \in \mathbb{N}$, then $x+y \in \mathbb{N}$ and $xy \in \mathbb{N}$.
(c) If X is a hereditary subset of $\mathbb{R}$ and $1 \in X$, then $\mathbb{N} \subset X$.

This axiom deserves a moment of meditation. Part (c) is intuitively clear: If X is hereditary and $1 \in X$, then $2 \in X$ since $2 = 1+1$, $3 \in X$ since $3 = 2+1$, etc. If you believe in mathematical induction you can actually *prove* (c). However, relying on mathematical induction goes against the rules of our game. By its very nature, induction deals with the positive integers, hence with real numbers. Then it must follow from the axioms and not have any a priori validity.

In fact, it is not hard to see that it really is a consequence of (c) of Axiom III. The Induction Principle can be formulated as follows:

Suppose for every $n \in \mathbb{N}$ a statement (or a formula) $P(n)$ is given. Suppose that

$$\left[\begin{array}{l} P(1) \text{ is true,} \\[4pt] \text{whenever } k \in \mathbb{N} \text{ and } P(k) \text{ is true, then so is } P(k+1). \end{array} \right. \qquad \text{(IP)}$$

Then $P(n)$ is true for every n in $\mathbb{N}$.

It is clear how one proves (IP): the set

$$X := \{n \in \mathbb{N} : P(n) \text{ is true}\}$$

contains 1 and is hereditary; hence, $\mathbb{N} \subset X$ by (c).

2.6
Now that we have Axiom III we can define the terms "integer" and "rational number," introduce the letters "$\mathbb{Z}$" and "$\mathbb{Q}$," and prove, for instance, that $\mathbb{N}$ is precisely the set of all positive integers. Also, we can define "sequence" and "limit" and prove the familiar theorems like

$$\lim_{n \to \infty} (x_n + y_n) = \lim_{n \to \infty} x_n + \lim_{n \to \infty} y_n.$$

However, we cannot prove that $\lim_{n\to\infty} n^{-1} = 0$. It can actually be shown that this formula is not implied by the axioms we have at this stage. We need:

Axiom IV (Archimedes-Eudoxus).
For every x in $\mathbb{R}$, there is an n in $\mathbb{N}$ with $x < n$.

Now we can show that $n^{-1} \to 0$. Indeed, take $\varepsilon > 0$. By the axiom, there is a $P \in \mathbb{N}$ with $\varepsilon^{-1} < P$. Then $|n^{-1}-0| = n^{-1} < \varepsilon$ as soon as $n \geq P$.

Another consequence of the Archimedes-Eudoxus axiom follows:

Theorem 2.7
Let $x \in \mathbb{R}$. Then there exists a unique integer in the interval $(x-1, x]$. (This integer is called the integer part *or* entire part *of x and denoted $[x]$.)*

Proof
Choose $M, N \in \mathbb{N}$ with $M > x$, $N > -x$; then $-N < x < M$. Consider the set

$$X := \{n \in \mathbb{N} : -N-1+n \leq x\}.$$

This X is a subset of $\mathbb{N}$, containing 1 but not equal to $\mathbb{N}$ itself; therefore, it cannot be hereditary. Thus, there is an n with $n \in X$, $n+1 \notin X$:

$$n \in \mathbb{N}, \quad -N-1+n \leq x, \quad -N+n > x.$$

Then, $-N-1+n$ is an integer in the interval $(x-1, x]$.

As for the uniqueness: If p and q are integers in $(x-1, x]$, then $|p-q| \in \mathbb{Z}$ and $|p-q| < 1$, so $|p-q| = 0$ and $p = q$. ∎

You probably knew the above theorem, if not its proof. The following result may be new for you.

Theorem 2.8

(i) *If $x \in \mathbb{R}$, then $x = \lim\limits_{n\to\infty} \frac{[nx]}{n}$. In particular, every number is the limit of a sequence of rational numbers.*
(ii) *Every interval contains a rational number.*

Proof

(i) For every n, we have $nx-1 < [nx] \leq nx$, so that $x-\frac{1}{n} < \frac{[nx]}{n} \leq x$. Hence, $\lim\limits_{n\to\infty} \frac{[nx]}{n} = x$, since $\lim\limits_{n\to\infty} \frac{1}{n} = 0$.
(ii) Every interval has a subinterval of the form (a, b) with $a, b \in \mathbb{R}$ and $a < b$. Set $c := \frac{a+b}{2}$. By Axiom IV, there is an n in $\mathbb{N}$ with $\frac{1}{n} < c-a$ and thereby $a < c-\frac{1}{n} < \frac{[nc]}{n} \leq c < b$, so $\frac{[nc]}{n} \in (a, b) \cap \mathbb{Q}$. ∎

2.9

Axioms I-IV give meaning to all the general concepts you have encountered in calculus: limit of a sequence, sum of a series, limit of a function, derivative, etc. We are not going to repeat all the definitions but will freely use the terms whenever we need them.

The axioms are still insufficient to establish the existence of certain specific objects such as the exponential function and the sine, or to prove theorems like l'Hospital's Rule and the Comparison Test for series. The time has come for our last axiom, the keystone in our building.

Axiom V (Dedekind).
If A and B are subsets of $\mathbb{R}$ such that

(a) $A \neq \varnothing$ and $B \neq \varnothing$,
(b) $A \cup B = \mathbb{R}$,
(c) for all $a \in A$ and $b \in B$: $a < b$,

then there is a number c for which

$$A = (-\infty, c), \ B = [c, \infty), \quad \text{or} \quad A = (-\infty, c], \ B = (c, \infty).$$

Think about it for a while and decide if you find it acceptable. (Newton might well have rejected it.)

2.10

Our axiom system is now finished. The whole body of calculus can be derived from it. We are not going to do so, our subject being topology, not analysis. There are, however, one or two theorems belonging to both disciplines and of those we will give precise proofs. Further on, in examples and comments, we will occasionally use other results from calculus. For their justifications we refer to texts on exact analysis.

Most applications of Dedekind's Axiom follow more easily from this less symmetrical variant:

Theorem 2.11 (Half-line Theorem)
Let L be a subset of $\mathbb{R}$ with the properties

(1) $L \neq \varnothing, L \neq \mathbb{R}$,
(2) *if $x \in L$ and $x' < x$, then $x' \in L$.*

Then there exists a number c with

$$(-\infty, c) \subset L \subset (-\infty, c].$$

Proof
$A := L$ and $B := \mathbb{R} \backslash L$ satisfy the requirements of Axiom V, so there is a c for which L is either $(-\infty, c)$ or $(-\infty, c]$.   ∎

2.12

A useful consequence of the Half-line Theorem is the "Least Upper Bound Theorem," probably familiar to you.

If X is a nonempty subset of $\mathbb{R}$ and if $b \in \mathbb{R}$, we call b an *upper bound* for X if $X \subset (-\infty, b]$, i.e., if

$$x \in X \quad \Longrightarrow \quad x \le b.$$

The upper bounds of the interval $(0, 1)$ are precisely the elements of $[1, \infty)$. The sets $\mathbb{N}$ and $\mathbb{R}$ have no upper bounds at all.

Theorem 2.13 (Least Upper Bound Theorem)

Let X be a nonempty subset of $\mathbb{R}$ that has upper bounds. Then among the upper bounds for X there is a smallest one.

Proof

Let L be the set of all numbers that are *not* upper bounds for X. Trivially, $L \neq \mathbb{R}$. For any $x \in X$, we have $x-1 \in L$, so $L \neq \varnothing$. It is obvious that L has Property (2), mentioned in the Half-Line Theorem. Hence, there is a number c with $L = (-\infty, c)$ or $L = (-\infty, c]$. In the first case, clearly c is the smallest upper bound. In the second case, the numbers $c+1, c+\frac{1}{2}, c+\frac{1}{3}, \ldots$ are upper bounds for X; then so is c itself. This means that the second case does not really occur. ∎

2.14

The least upper bound for a set X is also called the *supremum* of X, denoted

$$\sup X \qquad \text{or} \qquad \text{l.u.b. } X.$$

The sets $(0, 1)$, $[0, 1]$, and $\{1\}$ all have supremum 1. We see that $\sup X$ may or may not be an element of X. If it is, then it is the largest element of X. Conversely, if X has a largest element, then that is $\sup X$.

2.15

Let $X \subset \mathbb{R}$, $c \in \mathbb{R}$. We say that c is *adherent* to X if

$$[c-\varepsilon, c+\varepsilon] \cap X \neq \varnothing \qquad \text{for all } \varepsilon > 0.$$

Every element of X itself is adherent to X. Every real number is adherent to $\mathbb{Q}$, since $\mathbb{Q}$ intersects every interval [Theorem 2.8(ii)]. The number 0 is not adherent to $\mathbb{N}$, as $[-\frac{1}{2}, \frac{1}{2}] \cap \mathbb{N} = \varnothing$. It is, however, adherent to the interval $(0, 1)$.

If a set X has a supremum c, then c is adherent to X. Indeed, let $\varepsilon > 0$. Then $c-\varepsilon$ is not upper bound for X, so there is an $x \in X$ with $c-\varepsilon < x$. But c is an upper bound, from which $x \le c$. Now $x \in [c-\varepsilon, c+\varepsilon] \cap X$.

In 1.12, we have seen that continuity can be formulated in terms of convergent sequences. In the same vein we have:

Lemma 2.16

Let $X \subset \mathbb{R}$, $c \in \mathbb{R}$. Then c is adherent to X if and only if there is a sequence in X converging to c.

Proof

If c is adherent to X, then for every $n \in \mathbb{N}$, we can choose an x_n in $X \cap [c - \frac{1}{n}, c + \frac{1}{n}]$. Then $x_1, x_2, \ldots \in X$ and $x_n \to c$.

Conversely, if X contains a sequence $x_1, x_2, \ldots$ converging to c, then for every positive ε there is an n with $|x_n - c| \leq \varepsilon$; then $x_n \in [c - \varepsilon, c + \varepsilon] \cap X$.   ∎

Theorem 2.17 (Connectedness Theorem)

If A and B are nonempty subsets of $\mathbb{R}$ with $\mathbb{R} = A \cup B$, then there is a number that is adherent to both of them.

Proof

Take $a \in A$ and $b \in B$. Without restriction, suppose $a < b$. Set

$$X := \{x \in A : x \leq b\}.$$

X is nonempty ($a \in X$) and has b as an upper bound. Let c be its supremum. Then c is adherent to X, hence to A.

Take $\varepsilon > 0$. Obviously, $c + \varepsilon \notin X$. Thus, either $c + \varepsilon \notin A$ (so that $c + \varepsilon \in B$) or $c + \varepsilon > b$. In any case, $(c + \varepsilon) \wedge b$ lies in B and, of course, in $[c - \varepsilon, c + \varepsilon]$.

Thus, c is adherent to B.   ∎

For just this once, we return to analysis and show how the above can be used to prove a result that is well-known in calculus:

Theorem 2.18 (Intermediate Value Theorem)

Suppose f is a continuous function on an interval $[a,b]$. Let p be a number such that

$$f(a) \leq p \leq f(b) \quad or \quad f(a) \geq p \geq f(b).$$

Then p is a value of f, i.e., there is a c in $[a,b]$ with $f(c) = p$.

Proof

We extend f to a function g defined on all of $\mathbb{R}$ by

$$g(x) := \begin{cases} f(x) & \text{if } x \in [a, b], \\ f(a) & \text{if } x < a, \\ f(b) & \text{if } x > b. \end{cases}$$

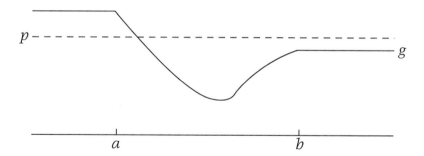

The function g is continuous everywhere and we are done if we can show that p is a value of g.

Applying the Connectedness Theorem to the sets $A := \{x \in \mathbb{R} : g(x) \geq p\}$ and $B := \{x \in \mathbb{R} : g(x) \leq p\}$, we obtain a number c that is adherent to both A and B. By Lemma 2.16, c is the limit of a sequence $x_1, x_2, \ldots$ of elements of A. Then $g(c) = \lim_{n \to \infty} g(x_n)$, whereas $g(x_n) \geq p$ for all n. Hence, $g(c) \geq p$. By the same token, $g(c) \leq p$ because c is adherent to B. Then $g(c) = p$.   ∎

It is this lemma that guarantees the existence of, e.g., $\sqrt{2}$. (See 2.5.) Taking $f(x) := x^2$ for $1 \leq x \leq 2$ and observing that $f(1) \leq 2 \leq f(2)$, we see that the equation $x^2 = 2$ must have a solution in $[1, 2]$. Observing that

$$0 \leq x \leq y \quad \Longrightarrow \quad x^2 \leq y^2,$$

we see that the equation has only one solution in $[1, 2]$, or even in $[0, \infty)$. *The* solution of the equation $x^2 = 2$ in $[0, \infty)$ is then denoted $\sqrt{2}$.

The following result is of interest both for topology and for analysis.

Theorem 2.19 (Cantor's Theorem)
Let

$$[a_1, b_1] \supset [a_2, b_2] \supset \cdots$$

be a sequence of intervals such that $\lim_{n \to \infty} b_n - a_n = 0$. Then there exists one and only one number c that lies in every $[a_n, b_n]$. (It follows that $c = \lim_{n \to \infty} a_n = \lim_{n \to \infty} b_n$.)

Proof
First, observe that for all n and m,

$$a_n \leq b_m,$$

since $a_n \leq a_{n+m} \leq b_{n+m} \leq b_m$. This means that every b_m is an upper bound for the set $\{a_1, a_2, \ldots\}$. By 2.13, this set has a least upper bound, c. Then $a_n \leq c$ for every n because c is an upper bound, and $c \leq b_m$ for every m because c is the *least* upper bound.   ∎

Extra: Axiom Systems

In principle, one can put together any group of statements, call them "axioms," and start proving theorems. Some axiom systems, however, turn out to be more satisfactory than others. There is nothing particularly sacred about the system we have presented, but it leads to an interesting and fruitful branch of mathematics.

Still, it is to some degree arbitrary. Analysis has been developed in order to obtain simple descriptions of natural phenomena, to perform certain calculations, and thereby to predict, say, the positions of the planets in the sky, or the motion of a ball rolling down a slope. It is based on our intuitive concept of real numbers. This concept has been formed in the course of history and there is no reason to assume that its evolution is finished. There have been times when only rational numbers were accepted as, in any sense, existing. Axiom V (Dedekind's Axiom) was formulated in 1872. This does not mean that only by that time its truth was discovered, but rather that in the course of the nineteenth century a choice was made - a fashion was set, you might say. It would have been possible to reject Axiom V and develop analysis in a different direction.[1] In fact, that has been done. There is the so-called "nonstandard Analysis" in which our Axioms I, II, and III are accepted, but the sequence $1, \frac{1}{2}, \frac{1}{3}, \ldots$ does not converge. This theory is just as "valid" as ours; it is simply different and, so far, less useful (although it does have its applications.)

Our choice of axioms is also arbitrary in another way.

Suppose a dissident mathematician reads through the preceding pages and finds Axioms I, II, III, and IV acceptable but rejects the Dedekind Axiom V. Out of curiosity, or amused by our absurdity, he reads on and finds the Connectedness Theorem perfectly in accordance with his own ideas. (After all, even such nonsense as Axiom V can lead to correct results.) He then decides to build his own theory, based on Axioms I, II, III, IV, and CT, our Connectedness Theorem.

We believe in his axioms. Then we must also believe in his conclusions, assuming his logic is correct. Thus, his theory will be part of ours.

It turns out that from his axiom system he can actually prove the statement we have called Axiom V. To see this, consider the following passage from his notebook.

> *Take sets A and B with the properties* (1), (2), (3) *mentioned in V. By my Axiom CT there is a number c adherent to A and B. There exist* $a_1, a_2, \ldots \in A$ *with* $a_n \to c$. *If* $b \in B$, *then (by* (3)) $a_n < b$ *for all n, hence* $c \leq b$. *Thus* $B \subset [c, \infty)$ *and therefore* $[-\infty, c) \subset A$. *Similarly,* $A \subset (-\infty, c]$ *and* $(c, \infty) \subset B$. *This proves* V.

---

[1] This is putting the cart before the horse. Historically, part of the theory came before the axiomatization.

It follows that *our* axioms are valid in *his* theory, and that, consequently, our entire theory will be contained in his.

We see that the two theories are really the same. Once one has accepted I, II, III, and IV as axioms, the choice between V and CT is just a matter of taste. In this sense, the axiom systems I-II-III-IV-V and I-II-III-IV-CT are "equivalent."

It can be shown that Axiom III [or rather, the statement that there *exists* a set $\mathbb{N} \subset \mathbb{R}$ with (a), (b) and (c)] and Axiom IV follow from Axioms I, II, and V. Thus, the axiom system I-II-V suffices for our theory and is equivalent to I-II-III-IV-V.

In the literature, various equivalent axiom systems for $\mathbb{R}$ are presented. Practically all consist of Axioms I and II, and a third axiom, occasionally with Axiom III and/or IV added. The most common combination is I-II-S, where S is the "Least Upper Bound Axiom," our Theorem 2.13. This S is a consequence of our axioms, so our analysis encompasses the I-II-S-theory. On the other hand, the books working with I-II-S invariably prove Axioms III and IV right away, while Axiom V is essentially a special case of S. Thus, our axiom system is equivalent to I-II-S.

Exercises

2.A. Prove that, for $x, y, z \in \mathbb{R}$:
   (a) If $x < y$ and $y \le z$, then $x < z$.
   (b) If $x < y$, then $-y < -x$.
   (c) $(-x)y = -(xy)$.
   [If you are unused to the axiomatic method, you may find it strange that formulas like these have to be proved. As in the case of (∗) of 2.3, the idea is not to make sure that they hold for the system of real numbers you have in mind but that they follow from the axioms. There are many mathematical structures satisfying the axioms; doing the exercise will establish the validity of (a)-(c) for all those structures.]

2.B. Give another proof of the identity $x0 = 0$ by starting with $x0 = x(0 + 0)$.

2.C. Let X be a nonempty subset of $\mathbb{R}$ that has upper bounds and therefore has a supremum. Let $Y \subset X$ and $Y \ne \varnothing$. Show that Y also has a supremum and that $\sup Y \le \sup X$.

2.D. Let X and Y be nonempty subsets of $\mathbb{R}$ such that

$$\sup X = 3 \quad \text{and} \quad \sup Y = 5.$$

What can you say about $\sup(X \cup Y)$? What about $\sup(X \cap Y)$?

2.E. Let X be a nonempty subset of $\mathbb{R}$. We call a number a a *lower bound* for X if

$$x \in X \quad \Longrightarrow \quad x \ge a.$$

(i) Prove: *If X has lower bounds, then among those there is a largest one.* The largest lower bound of X is called the *infimum* of X,

$$\inf X \quad\text{or}\quad \text{g.l.b. } X.$$

(ii) Let $Y := \{-x : x \in X\}$. Prove: *If X has lower bounds, then Y has upper bounds and*

$$\inf X = -(\sup Y).$$

2.F. The interval $(0, \infty)$ has no smallest element: if x is any element of $(0, \infty)$, then $\frac{1}{2}x$ is a smaller one. Thus, not every nonempty subset of $\mathbb{R}$ has a smallest element. However, prove that *every nonempty subset of $\mathbb{N}$ has a smallest element.* (Hint. Let $X \subset \mathbb{N}$, $X \neq \varnothing$. Deduce from Exercise 2.E that X has an infimum, c. Show that $c < [c]+1$, so that there must be an $x \in X$ with $c \leq x < [c]+1$. Prove that then $x \leq [c] \leq c$ and that, consequently, $c = x \in X$.)

2.G. Let $N \in \mathbb{N}$ be such that $\sqrt{N}$ is not an integer. Prove that then $\sqrt{N}$ is even irrational. (Hint. *Assume* $\sqrt{N} \in \mathbb{Q}$. Then the set

$$X := \{x \in \mathbb{N} : x\sqrt{N} \in \mathbb{N}\}$$

is nonempty. Show that, if $x \in X$ and $x' := x\sqrt{N} - x[\sqrt{N}]$, then $x' \in X$ and $x' < x$. Use Exercise 2.F to obtain a *contradiction*.)
   Thus, $\sqrt{2}, \sqrt{3}, \sqrt{5}, \ldots$ are irrational.

2.H. Show: *If $x \in \mathbb{R}$ is irrational and $a \in \mathbb{Q}$, $a \neq 0$, then $a+x$ and ax are irrational.*
   Show that *every interval contains an irrational number.* (Use the irrationality of $\sqrt{2}$, established in the previous exercise.)

2.I. Let

$$f(x) = a_0 + a_1 x + \cdots + a_n x^n \qquad (x \in \mathbb{R})$$

be a polynomial function, $a_n \neq 0$. Suppose n is odd. Show that the equation $f(x) = 0$ has at least one solution in $\mathbb{R}$.

2.J. Use the Intermediate Value Theorem to prove that if $f : [0, 1] \to [0, 1]$ is continuous, then there exists an x in $[0, 1]$ with $f(x) = x$.

2.K. Let $x_1, x_2, \ldots$ be real numbers such that

$$|x_n - x_m| \leq \left| \frac{1}{n} - \frac{1}{m} \right| \quad \text{for all } n, m.$$

Show that the sequence $(x_n)_{n \in \mathbb{N}}$ converges. (Hint. Apply Cantor's Theorem with $a_n = x_n - \frac{1}{n}$ and $b_n = x_n + \frac{1}{n}$.)

3 CHAPTER

Convergent Sequences and Continuity

Subsequences

3.1

For a good understanding of this chapter, one has to keep in mind the formal definition of a *sequence* as a function whose domain is $\mathbb{N}$. This is not going to prevent us from using the notations $x_1, x_2, \ldots$ and $(x_n)_{n \in \mathbb{N}}$.

From a sequence $x_1, x_2, \ldots$, one obtains a "subsequence" roughly by deleting terms. The main rule is that infinitely many terms must be left: A subsequence of $x_1, x_2, \ldots$ is a sequence in its own right. (A precise definition follows in 3.2.)

For instance, the sequence

$$1, 1, 2, 2, 3, 3, 4, 4, \ldots$$

has subsequences

$$1, 2, 3, 4, 5, 6, 7, \ldots$$

and

$$1, 1, 3, 3, 5, 5, 7, \ldots.$$

It is against the rules to change the order of the terms or introduce new repetitions: $2, 1, 4, 3, 6, 5, \ldots$ and $1, 1, 1, 1, 1, \ldots$ are not subsequences of the sequence $1, 1, 2, 2, 3, 3, \ldots$.

3.2
If we start out with a sequence of numbers (or other objects)

$$x_1, x_2, x_3, x_4, x_5, x_6, x_7, x_8, \ldots$$

and delete a few terms, then we have left a sequence

$$x_p, x_q, x_r, \ldots$$

with $p < q < r < \cdots$.

This observation brings us to the formal definition.

A map $\alpha : \mathbb{N} \to \mathbb{N}$ (a sequence of positive integers) is called *strictly increasing* if

$$\alpha(1) < \alpha(2) < \alpha(3) < \cdots.$$

A sequence $y_1, y_2, \ldots$ is said to be a *subsequence* of a sequence $x_1, x_2, \ldots$ if there exists a strictly increasing $\alpha : \mathbb{N} \to \mathbb{N}$ with

$$y_n = x_{\alpha(n)} \text{ for all } n \text{ in } \mathbb{N},$$

i.e. if $y_1, y_2, \ldots$, considered as a function $\mathbb{N} \to \mathbb{R}$, is the composition of $x_1, x_2, \ldots$ with some strictly increasing $\alpha : \mathbb{N} \to \mathbb{N}$.

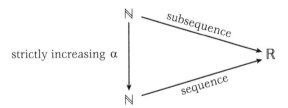

For instance, a "tail" $x_N, x_{N+1}, x_{N+2}, \ldots$ (see 1.9) is a subsequence of $x_1, x_2, \ldots$ obtained from $\alpha(n) = n+N-1$ $(n \in \mathbb{N})$.

Formally, "the sequence $1, 2, 3, \ldots$" is the function $n \mapsto n$ $(n \in \mathbb{N})$, and the strictly increasing maps $\mathbb{N} \to \mathbb{N}$ are precisely its subsequences.

Two simple remarks:

Lemma 3.3
If $y_1, y_2, \ldots$ is a subsequence of $x_1, x_2, \ldots$ and $z_1, z_2, \ldots$ is a subsequence of $y_1, y_2, \ldots$, then $z_1, z_2, \ldots$ is a subsequence of $x_1, x_2, \ldots$.

Proof
There exist strictly increasing $\alpha : \mathbb{N} \to \mathbb{N}$ and $\beta : \mathbb{N} \to \mathbb{N}$ with $y_n = x_{\alpha(n)}$ and $z_m = y_{\beta(m)}$ for all n, m. Then $z_m = x_{\alpha(\beta(m))} = x_{(\alpha \circ \beta)(m)}$ for all m, and $\alpha \circ \beta : \mathbb{N} \to \mathbb{N}$ is strictly increasing. ∎

Lemma 3.4
If $\alpha : \mathbb{N} \to \mathbb{N}$ is strictly increasing, then $\alpha(n) \geq n$ for all n.

Proof

By induction.   ∎

3.5

One should take care not to confuse a sequence

$$x_1, x_2, x_3, \ldots$$

with the set of its entries

$$\{x_1, x_2, x_3, \ldots\}.$$

A sequence always has infinite length, so to speak, but the set of the entries may well be finite. Also, by changing the order of the terms, one obtains a different sequence but with the same entries. (Still, we will occasionally sin and speak of, e.g., a "sequence lying in [0, 1].")

There is one situation where making the distinction verges on pedantry. Theoretically, the expression "the sequence of all prime numbers" is meaningless: The prime numbers form a set, not a function on $\mathbb{N}$. However, any reasonable person will understand that the sequence

$$2, 3, 5, 7, 11, 13, \ldots$$

is meant and not, for instance,

$$3, 3, 3, 2, 11, 37, 3, \ldots.$$

More generally, without danger of confusion, one may identify the infinite subsets of $\mathbb{N}$ with the strictly increasing sequences of positive integers.

Theorem 3.6

If $\lim_{n\to\infty} x_n = a$, then all subsequences of $x_1, x_2, \ldots$ converge to a.

Proof

(using the definition of convergence): Let $y_1, y_2, \ldots$ be a subsequence of $x_1, x_2, \ldots$. Take a strictly increasing $\alpha : \mathbb{N} \to \mathbb{N}$ with $y_n = x_{\alpha(n)}$ for all n.

Let $\varepsilon > 0$. There is an $N \in \mathbb{N}$ with

$$n \geq N \quad \Longrightarrow \quad |x_n - a| < \varepsilon.$$

For all $m \geq N$, we have $\alpha(m) \geq N$ by the previous lemma, so that $|x_{\alpha(m)} - a| < \varepsilon$, i.e., $|y_m - a| < \varepsilon$.

Thus, $y_m \to a$.   ∎

3.7

A function f is said to be *bounded* if there exists a number K such that all values of f lie in the interval $[-K, K]$. Thus, the sine function on $\mathbb{R}$ is bounded; the exponential function is not (but its restriction to [0, 20] is).

This definition is applicable to functions defined on arbitrary sets. In particular, a sequence $x_1, x_2, \ldots$ is bounded if there exists a number K such that

$$|x_n| \leq K \quad \text{for all } n.$$

Every convergent sequence is bounded. Indeed, let $x_n \to a$. There is an N such that $|x_n - a| < 1$ for all $n \geq N$. Then $|x_n| \leq |a| + 1$ as soon as $n \geq N$, and

$$|x_n| \leq |x_1| + |x_2| + \cdots + |x_{N-1}| + |a| + 1$$

for *all* n.

It is, of course, not true that every bounded sequence converges. However, we do have the following important fact.

Theorem 3.8 (Bolzano-Weierstrass Theorem)
Every bounded sequence in $\mathbb{R}$ has a convergent subsequence.

Proof
Let $(x_n)_{n \in \mathbb{N}}$ be a sequence in $\mathbb{R}$ and let $M \in \mathbb{R}$ be such that $|x_n| < M$ for all n.

Consider the subset L of $\mathbb{R}$ defined by

$$y \in L \quad \Longleftrightarrow \quad y \leq x_n \quad \text{for infinitely many values of } n.$$

If $y \in L$ and $y' < y$, then $y' \in L$. Furthermore, $-M \in L$ and $M \notin L$. Hence, by the Half-line Theorem, there is a number c such that

$$(-\infty, c) \subset L \subset (-\infty, c].$$

We construct a subsequence converging to c.

Take $\varepsilon > 0$. Then $c - \varepsilon \in L$, so there are infinitely many values of n for which $c - \varepsilon \leq x_n$. But $c + \varepsilon \notin L$, so we can have $c + \varepsilon \leq x_n$ for only finitely many of those values. This implies that there are infinitely many values of n with $c - \varepsilon \leq x_n < c + \varepsilon$.

Now we make the subsequence. First, choose any $\alpha(1) \in \mathbb{N}$ for which

$$c - 1 \leq x_{\alpha(1)} < c + 1.$$

There are infinitely many values of n with $c - \frac{1}{2} \leq x_n < c + \frac{1}{2}$. Only finitely many of them can be $\leq \alpha(1)$, so we can choose an $\alpha(2) \in \mathbb{N}$ for which

$$c - \tfrac{1}{2} \leq x_{\alpha(2)} < c + \tfrac{1}{2} \quad \text{and} \quad \alpha(2) > \alpha(1).$$

Similarly, we can find an $\alpha(3)$ in $\mathbb{N}$ such that

$$c - \tfrac{1}{3} \leq x_{\alpha(3)} < c + \tfrac{1}{3} \quad \text{and} \quad \alpha(3) > \alpha(2),$$

and so on. We obtain a strictly increasing $\alpha : \mathbb{N} \to \mathbb{N}$ and $x_{\alpha(n)} \to c$. ∎

The Bolzano-Weierstrass Theorem is a very useful tool in analysis. (Indeed, assuming Axioms I and II of §2, it is equivalent to Axiom V.) We apply it to prove a result you know from calculus.

Theorem 3.9

Let D be a closed bounded interval and let $f:D \to \mathbb{R}$ be continuous. Then f attains a largest value, i.e., there is a c in D with

$$f(c) \geq f(x) \quad \text{for all } x \text{ in } D.$$

(Similarly, f attains a smallest value.)

Proof

(I) First, assume f is bounded. Then the nonempty set $\{f(x) : x \in D\}$ has upper bounds; let s be its supremum:

$$s := \sup\{f(x) : x \in D\}.$$

Obviously, $s \geq f(x)$ for all $x \in D$; we are done if s is itself a function value.

For every $n \in \mathbb{N}$, $s-n^{-1}$ is smaller than s and hence $s-n^{-1}$ is no upper bound of $\{f(x) : x \in D\}$. Then there is an x_n in D with $s-n^{-1} < f(x_n)$. But, of course, $f(x_n) \leq s$, so

$$f(x_n) \to s.$$

The sequence $x_1, x_2, \ldots$ has a subsequence $x_{\alpha(1)}, x_{\alpha(2)}, \ldots$ converging to some number c in D. Then

$$f(x_{\alpha(n)}) \to f(c).$$

But the sequence $f(x_{\alpha(1)}), f(x_{\alpha(2)}), \ldots$ is a subsequence of $f(x_1), f(x_2), \ldots$ and hence converges to s (Theorem 3.6). Thus, $s = f(c)$ and we are done.

(II) We have proved the theorem for bounded functions. It is time to drop that restriction: Let f be any continuous function on D. Define $g : D \to \mathbb{R}$ by

$$g(x) := \tan^{-1} f(x) \qquad (x \in D).$$

Then g is continuous and bounded. Hence, there is a c in D with $g(c) \geq g(x)$ for all x. Then $f(c) \geq f(x)$ for all x. ∎

In a sense, the second part of the above proof is void:

Corollary 3.10

On a closed bounded interval, every continuous function is bounded.

Proof

Let $f : D \to \mathbb{R}$ be continuous, where D is a closed bounded interval. Apply the previous theorem to the function $x \mapsto |f(x)|$. ∎

Uniform Continuity

3.11

Consider the following functions with domain $(0, \infty)$:

$$\left.\begin{array}{l} f(x) := x \\ g(x) := \sqrt{x} \\ h(x) := x^2 \end{array}\right] \quad (x > 0).$$

We know that all three are continuous, but let us compare their continuity proofs. "Continuous" here means "continuous at every point of $(0, \infty)$," so in each case, we have to take $a \in (0, \infty)$ and $\varepsilon > 0$ and find a suitable δ.

For f, we choose $\delta := \varepsilon$. If $x \in (0, \infty)$ and $|x-a| < \delta$, then

$$|f(x)-f(a)| = |x-a| < \delta = \varepsilon.$$

For g, we choose $\delta := \varepsilon\sqrt{a}$. If $x \in (0, \infty)$ and $|x-a| < \delta$, then

$$|g(x)-g(a)| = |\sqrt{x}-\sqrt{a}|$$

$$= \frac{|x-a|}{\sqrt{x}+\sqrt{a}} < \frac{\delta}{0+\sqrt{a}} = \varepsilon.$$

For h, we choose $\delta := \sqrt{a^2+\varepsilon} - a$. If $x \in (0, \infty)$ and $|x-a| < \delta$, then

$$|h(x)-h(a)| = |x-a|\,|x+a|$$

$$< \delta(2a+\delta) = \cdots = \varepsilon.$$

In the above, our choices for δ were ε, $\varepsilon\sqrt{a}$, and $\sqrt{a^2+\varepsilon}-a$, respectively. You may notice that in the first case our δ was independent of a, in the other two cases it was not. *Could* we have obtained a δ that was the same for all a?

The answer is affirmative in the case of g. Taking $\delta := \varepsilon^2$, for all $x \in (0, \infty)$ with $|x-a| < \delta$ we have

$$g(x) = \sqrt{x} < \sqrt{a+\delta} \le \sqrt{a} + \sqrt{\delta} = g(a)+\varepsilon,$$

$$g(a) = \sqrt{a} < \sqrt{x+\delta} \le \sqrt{x} + \sqrt{\delta} = g(x)+\varepsilon,$$

and, therefore, $|g(x)-g(a)| < \varepsilon$.

For the function h, however, we can see that it is impossible to find a δ that does not depend on a. Indeed, whatever positive number δ we try, taking $a := \varepsilon\delta^{-1}$ and $x := a+\frac{1}{2}\delta$ we always have $x \in (0, \infty)$ and $|x-a| < \delta$, but

$$h(x)-h(a) = (a+\tfrac{1}{2}\delta)^2 - a^2 > a\delta = \varepsilon.$$

We see that in this respect the behavior of h differs from that of f and g. We will say that f and g are "uniformly continuous" but h is not.

3.12
Let f be a function on a subset D of $\mathbb{R}$. We call f *uniformly continuous* if

for every $\varepsilon > 0$ there exists a $\delta > 0$ such that

$$x, y \in D, \; |x-y| < \delta \quad \Longrightarrow \quad |f(x)-f(y)| < \varepsilon.$$

Think about this definition and check that the functions f and g of 3.11 are indeed uniformly continuous whereas h is not.

Every uniformly continuous function is continuous; apparently, the converse is false. However,

Theorem 3.13
On a closed bounded interval, every continuous function is uniformly continuous.

Proof
Let D be a closed bounded interval and let $f : D \to \mathbb{R}$ be continuous. Take $\varepsilon > 0$; we need a positive δ such that

$$x, y \in D, \; |x-y| < \delta \quad \Longrightarrow \quad |f(x)-f(y)| < \varepsilon.$$

Suppose such a δ does not exist. Then, whatever positive δ we try, there will always be x and y in D with

$$|x-y| < \delta \quad \text{but} \quad |f(x)-f(y)| \geq \varepsilon.$$

In particular we try $\delta = 1, \delta = \frac{1}{2}, \delta = \frac{1}{3}, \dots$. We see that for every $n \in \mathbb{N}$, there exist x_n and y_n for which

$$|x_n-y_n| < \frac{1}{n} \quad \text{but} \quad |f(x_n)-f(y_n)| \geq \varepsilon.$$

The sequence $x_1, x_2, \dots$ has a subsequence $x_{\alpha(1)}, x_{\alpha(2)}, \dots$ converging to a number c in D. For each n,

$$|x_{\alpha(n)}-y_{\alpha(n)}| < \frac{1}{n} \leq \frac{1}{\alpha(n)},$$

so $y_{\alpha(n)} \to c$. As f is continuous at c, we must have $f(x_{\alpha(n)}) \to f(c)$ and $f(y_{\alpha(n)}) \to f(c)$, *contradicting* the fact that $|f(x_{\alpha(n)})-f(y_{\alpha(n)})| \geq \varepsilon$ for all n. ∎

The Plane

All that we have been doing so far might well be classified as analysis. Our first step in the direction of topology consists of extending part of the above to functions of several variables. We restrict ourselves to two variables, but you may find it worthwhile to keep an eye on further generalization.

3.14
The coordinates of a vector x in $\mathbb{R}^2$ will be denoted $(x)_1$ and $(x)_2$:

$$x = \big((x)_1, (x)_2\big).$$

The small parentheses are needed because we will often work with sequences of vectors $x_1, x_2, \ldots$ and we wish to avoid confusion of the initial element of such a sequence and the first coordinate of a vector x. Admittedly, the parentheses will clutter up our formulas, but we will use coordinates sparingly.

As you know, for $x \in \mathbb{R}^2$ one defines the *length* of x to be the number

$$\|x\| := \sqrt{(x)_1^2 + (x)_2^2}.$$

The (Cartesian) *distance* between two vectors, x and y, is

$$\|x-y\|,$$

the length of the difference $x-y$.

3.15
We say that a sequence $(x_n)_{n\in\mathbb{N}}$ in $\mathbb{R}^2$ *converges to* an element x of $\mathbb{R}^2$,

$$x_n \to x,$$

if

for every $\varepsilon > 0$, there exists an $N \in \mathbb{N}$ such that

$$n \geq N \quad \Longrightarrow \quad \|x_n-x\| < \varepsilon,$$

which means precisely the same as

$$\lim_{n\to\infty} \|x_n-x\| = 0. \tag{$*$}$$

Let $(x_n)_{n\in\mathbb{N}}$ be a sequence in $\mathbb{R}^2$ and let $x \in \mathbb{R}^2$. For every n in $\mathbb{N}$ we have

$$|(x_n)_1-(x)_1| \leq \|x_n-x\|.$$

Hence, if $(*)$ is true, then $(x_n)_1 \to (x)_1$ and, similarly, $(x_n)_2 \to (x)_2$.
The converse is also true, since for all n

$$\|x_n-x\| = \sqrt{\big((x_n)_1-(x)_1\big)^2 + \big((x_n)_2-(x)_2\big)^2}.$$

Thus:

Theorem 3.16
Let $x, x_1, x_2, \ldots \in \mathbb{R}^2$. Then,

$$x_n \to x \quad \Longleftrightarrow \quad (x_n)_1 \to (x)_1 \quad and \quad (x_n)_2 \to (x)_2.$$

As a byproduct of this formula we obtain uniqueness of the limit: If $x_n \to x$ and $x_n \to y$, then $x = y$. This allows us to use the notation

$$\lim_{n \to \infty} x_n.$$

There are some easily verifiable rules, such as

$$\text{if } x_n \to x \text{ and } y_n \to y, \text{ then } x_n + y_n \to x + y.$$

3.17
Given a sequence $x_1, x_2, \ldots$ of vectors, we define its *subsequences* to be all sequences of the form $x_{\alpha(1)}, x_{\alpha(2)}, \ldots$, where $\alpha(1), \alpha(2), \ldots \in \mathbb{N}$ and $\alpha(1) < \alpha(2) < \cdots$. Precisely as in 3.5: *If a sequence $(x_n)_{n \in \mathbb{N}}$ converges to x, then so do all of its subsequences.*

A sequence $(x_n)_{n \in \mathbb{N}}$ of vectors is *bounded* if there exists a number M such that

$$\|x_n\| \leq M \quad \text{for all } n.$$

This is the case if and only if the coordinate sequences $(x_1)_1, (x_2)_1, (x_3)_1, \ldots$ and $(x_1)_2, (x_2)_2, (x_3)_2, \ldots$ are both bounded.

With Theorem 3.16, it is perfectly easy to see that every converging sequence is bounded. We also have a two-dimensional analog of the Bolzano-Weierstrass Theorem 3.8:

Theorem 3.18
In $\mathbb{R}^2$, every bounded sequence has a convergent subsequence.

Proof
Let $x_1, x_2, \ldots$ be a bounded sequence in $\mathbb{R}^2$.

Then the sequence of first coordinates $(x_1)_1, (x_2)_1, \ldots$ is bounded in $\mathbb{R}$ and, hence, has a convergent subsequence. The reader is asked to verify that this means that the sequence $x_1, x_2, \ldots$ has a subsequence $y_1, y_2, \ldots$ for which

$$\lim_{n \to \infty} (y_n)_1 \quad \text{exists.} \tag{*}$$

The number sequence $(y_1)_2, (y_2)_2, (y_3)_2, \ldots$ is bounded and therefore has a convergent subsequence. Then the sequence of vectors $y_1, y_2, y_3, \ldots$

has a subsequence $z_1, z_2, \ldots$ such that

$$\lim_{n \to \infty} (z_n)_2 \quad \text{exists.}$$

But by $(*)$

$$\lim_{n \to \infty} (z_n)_1 \quad \text{exists.}$$

Thus, the sequence $z_1, z_2, \ldots$ (a subsequence of $x_1, x_2, \ldots$) converges. ∎

3.19
Let f be a function whose domain is a subset D of $\mathbb{R}^2$ and let $a \in D$. We say that f is *continuous* at a if

for every $\varepsilon > 0$, there exists a $\delta > 0$ such that

$$x \in D, \ \|x{-}a\| < \delta \quad \Longrightarrow \quad |f(x){-}f(a)| < \varepsilon.$$

This definition is in accordance with the one you know from calculus, except for the fact that most calculus books require a priori that D contain a disk centered at a.

As a result of our greater generality, for certain combinations of D and a continuity may be vacuous. If, say, $D = \mathbb{Z} \times \mathbb{Z}$ and $a = (3, 5)$, then every function on D is continuous at a. Indeed, whatever function f and whatever positive ε we take, with $\delta = \frac{1}{2}$ we necessarily have

$$x \in D, \ \|x{-}a\| < \delta \quad \Longrightarrow \quad x = a \quad \Longrightarrow \quad |f(x){-}f(a)| < \varepsilon.$$

[This is the same phenomenon we have already observed in a slightly different context in 1.8(ii).]

The following theorem is perfectly analogous to Theorem 1.12 and its proof is nothing new either.

Theorem 3.20
Let $D \subset \mathbb{R}^2$, $f{:}D \to \mathbb{R}$, and $a \in D$. Then the conditions (α) and (β) are equivalent.

(α) *f is continuous at a.*
(β) *For every sequence $(x_n)_{n \in \mathbb{N}}$ in D with $x_n \to a$, we have $f(x_n) \to f(a)$.*

Proof
$(\alpha) \Longrightarrow (\beta)$. Let $x_1, x_2, \ldots \in D$, $x_n \to a$; let $\varepsilon > 0$. There exists a $\delta > 0$ such that $|f(x){-}f(a)| < \varepsilon$ for all $x \in D$ for which $\|x{-}a\| < \delta$. There exists an $N \in \mathbb{N}$ such that $\|x_n{-}a\| < \delta$ for all n with $n \geq N$. Then $|f(x_n){-}f(a)| < \varepsilon$ as soon as $n \geq N$.

$(\beta) \Longrightarrow (\alpha)$ Let $\varepsilon > 0$. We need a positive number δ with

$$x \in D, \ \|x{-}a\| < \delta \quad \Longrightarrow \quad |f(x){-}f(a)| < \varepsilon.$$

Suppose such a δ does not exist. Then, for every $\delta > 0$, there exists an x in D such that

$$\|x - a\| < \delta \quad \text{but} \quad |f(x) - f(a)| \geq \varepsilon.$$

In particular, taking $\delta = 1$, $\delta = \frac{1}{2}$, $\delta = \frac{1}{3}, \ldots$ we see that there exist $x_1, x_2, \ldots$ in D with, for every n,

$$\|x_n - a\| < \frac{1}{n} \quad \text{but} \quad |f(x_n) - f(a)| \geq \varepsilon.$$

Thus, there is a sequence $(x_n)_{n \in \mathbb{N}}$ in D for which $x_n \to a$ but not $f(x_n) \to f(a)$. *Contradiction.*   ∎

3.21

From this theorem we see that, for instance, the product of two continuous functions is continuous. Indeed, let $D \subset \mathbb{R}^2$, $a \in D$ and let f and g be functions on D, both continuous at a. Define $h : D \to \mathbb{R}$ by

$$h(x) := f(x)g(x) \qquad (x \in D).$$

We claim that h is continuous at a. (This will not surprise you, but the proof is of more interest than the statement itself.) To substantiate our claim, we consider a sequence $(x_n)_{n \in \mathbb{N}}$ in D that converges to a. We wish to prove $h(x_n) \to h(a)$. But that is a simple consequence of the rules for converging sequences we have listed in Theorem 1.11:

$$h(x_n) = f(x_n)g(x_n) \to f(a)g(a) = h(a).$$

You will see that, in precisely the same way, one can show that if $f : D \to \mathbb{R}$ is continuous at a, then so are, for instance, the functions

$$x \mapsto e^{f(x)} \qquad (x \in D)$$

and [if $f(x) \neq 0$ for all x]

$$x \mapsto \frac{1}{f(x)} \qquad (x \in D).$$

3.22

A function f on a subset D of $\mathbb{R}^2$ is called *continuous* if it is continuous at every point of D. Occasionally, we call $f : D \to \mathbb{R}$ *sequentially continuous* if

$$\left. \begin{array}{l} a, x_1, x_2, \ldots \in D \\ x_n \to a \end{array} \right] \quad \Longrightarrow \quad f(x_n) \to f(a).$$

Apparently, continuity is the same as sequential continuity, and you may wonder why we introduce two names for one concept. Bear with us for a while. In Part III you will see our reasons.

3.23

At this stage, we do not intend to set up an entire theory of continuous functions on subsets of $\mathbb{R}^2$. We only wish to make two remarks, in the spirit of Theorems 3.9 and 3.13.

In Theorem 3.9 we have proved that on a closed bounded interval D every continuous function has a largest value. If you carefully read over the proof (and the authors urge you to do so) you will notice that all we had to know about D was

$$\left.\begin{array}{l} \text{every sequence in } D \text{ has a subsequence} \\[2mm] \text{converging to an element of } D. \end{array}\right] \qquad \text{(SC)}$$

This means that the proof does not work only for closed bounded intervals but for every set D that has the property (SC), called *sequential compactness*.

It is really better than that. The proof of Theorem 3.9 does not really use the fact that D is a subset of $\mathbb{R}$: It works just as well for "sequentially compact" subsets of $\mathbb{R}^2$. (Check it; do not trust us.)

One thing and the other lead to

Theorem 3.24

Let D be a sequentially compact subset of $\mathbb{R}^2$; let $f:D \to \mathbb{R}$ be continuous. Then f attains a largest value.

(And a smallest. Hence, f is bounded.)

The same reasoning applies to Theorem 3.13 on uniform continuity; its proof works for any sequentially compact subset of $\mathbb{R}$ or $\mathbb{R}^2$. (All one really needs is a definition of uniform continuity for functions on subsets of $\mathbb{R}^2$. It is left to the reader to provide one.)

Theorem 3.25

On a sequentially compact subset of $\mathbb{R}^2$, every continuous function is uniformly continuous.

All this is vacuous as long as we do not have explicit examples of sequentially compact sets in $\mathbb{R}^2$. Fortunately, they are easy to find.

Examples 3.26

(i) Let D be the "closed unit square" $[-1, 1] \times [-1, 1]$:

$$D := \{x \in \mathbb{R}^2 : (x)_1 \in [-1, 1], (x)_2 \in [-1, 1]\}.$$

Take a sequence $x_1, x_2, \ldots$ in D. It follows from Theorem 3.18 that there is a subsequence $y_1, y_2, \ldots$, converging to some $y \in \mathbb{R}^2$. Is y an element of D? Of course it is: Its coordinates are limits of sequences in $[-1, 1]$ and therefore lie in $[-1, 1]$ themselves.

(ii) Let D be the "closed unit disk"

$$\{x \in \mathbb{R}^2 : \|x\| \le 1\}.$$

Again, any sequence $x_1, x_2, \ldots$ in D has a subsequence $y_1, y_2, \ldots$ that converges to some y in $\mathbb{R}^2$. For the sequential compactness of D, we need $\|y\| \le 1$. But $\|y\|^2 = (y)_1^2 + (y)_2^2 = \lim_{n \to \infty}(y_n)_1^2 + (y_n)_2^2 = \lim_{n \to \infty} \|y_n\|^2 \le 1$.

Extra: Bolzano (1781-1848)

The history of mathematics reveals some oddities. Bernhard Bolzano's place in the development of analysis is one of them. If due credit were given by naming theorems and definitions after their discoverers, Cauchy sequences (see 7.2) would be called Bolzano sequences and Theorem 4.28 would be cited as the Bolzano Closed Curve Theorem. Introductory chapters to analysis textbooks would rename Dedekind's Axiom (our Theorem 2.13) into Bolzano's Axiom. Weierstrass' name would be dropped from the Bolzano-Weierstrass Theorem and Theorem 2.18 would be known as Bolzano's Intermediate Value Theorem. Furthermore, Bolzano would be credited for introducing the modern definitions of a convergent sequence, a convergent series, continuity, and differentiability. And there is much more. His thoughts on infinite sets bridge the period from Galileo to Cantor (see Chapter 19), he constructed the set of real numbers, was the first to recognize the need for an *axiom* such as our Axiom V and he gave an explicit example of a continuous function that is nowhere increasing or decreasing. (In Exercise 8.I we prove the existence of such a function.)

Bolzano was born and lived in the city of Prague in Bohemia, at the time part of the Austrian-Hungarian Empire. The first half of the nineteenth century was a period of political instability in Europe, starting with the French Revolution and the Napoleonic Wars. Bohemia was no exception. There was much resistance against the Austrian regime and constant striving to obtain a higher level of autonomy, if not complete independence. The atmosphere was hardly conducive to scholarly pursuits.

Bolzano was no recluse. He studied Philosophy, Theology, and Mathematics. After his doctorate (his thesis was on geometry) in 1805, he chose to be a Catholic priest and was appointed a theology professor in Prague. His desire for clarity and his sympathy for socialist ideas made him popular with the Praguers and unpopular with his superiors. In 1819, he was fired because of his political activities and afterward he was accused of heresy. Neither this nor his bad health kept him from being active. For the rest of his life he was financially supported by friends, but between

1817 and 1840 he had no opportunity to publish his mathematical and philosophical ideas.

For mathematics, the age was also one of unrest. During the preceding century, the basic results of calculus had become well known, but there were no precise definitions and not much of the careful reasoning we are accustomed. As a consequence, there were controversies about interpretations of certain mathematical statements. Bolzano was the first to provide modern rigor. He went through an enormous amount of work. His total output (not all of a mathematical nature) is expected to fill 56 volumes.

The letdown is that his ideas went unnoticed for half a century. Everything was discovered independently, but later, by others, who have gotten the limelight. Most of the credit has gone to Bolzano's contemporary, the Frenchman Augustin-Louis Cauchy (1789-1857). It is not without interest to compare the two: Bolzano, living in isolation, ignored by the world, and Cauchy, one of the world's most famous scholars, who for his merits was made a baron by King Charles X. Without disparaging Cauchy, we must say that Bolzano introduced exactness before Cauchy did, and, unlike Cauchy, strictly adhered to it. (Also, he seems to have been a nicer person to have around.)

Further Reading

Russ, S., Bolzano's Analytic Programme, *Mathematical Intelligencer* 14, 1992, 45-53.

Exercises

3.A. Let $f : [a, b] \to \mathbb{R}$ be a continuous function. Show that there exists a w in $[a, b]$ such that

$$\int_a^b f(t)dt = f(w)(b-a).$$

(Hint. f attains a smallest value A and a largest value B. Show that $A \leq \frac{1}{b-a} \int_a^b f(t)dt \leq B$.)

3.B. Let $(x_n)_{n \in \mathbb{N}}$ be a bounded sequence in $\mathbb{R}$ that does not converge to 0. Show that there must exist a subsequence that converges to some number different from 0. (Of course, there may also be subsequences with limit 0.)

3.C. Let J be an interval and f a differentiable function on J whose derivative is bounded. Show that f is uniformly continuous (Use the Mean Value Theorem).

Observe that boundedness of f' is not *necessary* for uniform continuity of f. Indeed, the function $x \longmapsto \sqrt{x}$ ($x > 0$) is uniformly continuous and differentiable, but its derivative is *not* bounded.

3.D. The question of whether the function $x \longmapsto x^2$ is uniformly continuous is, strictly speaking, meaningless as long as we do not specify the domain of the function. For example, by Theorem 3.13,

$$x \longmapsto x^2 \qquad (x \in [1, 2])$$

is uniformly continuous, but from Corollary 3.10 we know that

$$x \longmapsto x^2 \qquad \left(x \in (0, \infty) \right)$$

is not. In this respect, uniform continuity is like boundedness (the logarithm is bounded on $[1, 2]$ but not on $[1, \infty)$) and injectivity ($x \longmapsto x^2$ is injective on $[0, \infty)$ but not on $\mathbb{R}$).

Which of the following functions are (or is) uniformly continuous? You may have use here for the result of the previous exercise.
- (a) $x \longmapsto \log x$ $\quad$ ($x \in [1, \infty)$).
- (b) $x \longmapsto \log x$ $\quad$ ($x \in (0, 1)$).
- (c) $x \longmapsto \sin x$ $\quad$ ($x \in \mathbb{R}$).
- (d) $x \longmapsto \sin \sqrt{x}$ $\quad$ ($x > 1$).
- (e) $x \longmapsto \cos x^{-1}$ $\quad$ ($0 < x \le 1$).
- (f) $x \longmapsto (1+x)^{-1}$ $\quad$ ($0 < x \le 1$).

3.E. Show that the subset

$$\{0\} \cup \{1, \tfrac{1}{2}, \tfrac{1}{3}, \ldots\}$$

of $\mathbb{R}$ is sequentially compact.

3.F. Let $a, b, c \in \mathbb{R}^2$. For a nonnegative number r, we say that a, b, and c lie in a disk with radius r if there exists a point x of $\mathbb{R}^2$ such that

$$\|a-x\| \le r, \quad \|b-x\| \le r, \text{ and } \|c-x\| \le r.$$

(Here we allow a disk to have radius 0.) Show that the set

$$\{r \in [0, \infty) : a, b, c \text{ lie in a disk with radius } r\}$$

has a smallest element. (Hint. Let r_0 be the infimum of the set. Show that r_0 is adherent to the set and apply Theorem 3.18.)

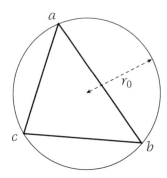
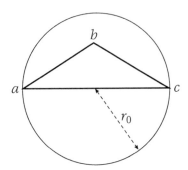

3.G. Let $\Gamma := \{(x, y) \in \mathbb{R}^2 : x^2+y^2 = 1\}$ and let $f : \Gamma \to \mathbb{R}$ be continuous. Prove the existence of a point p of Γ for which $f(p) = f(-p)$. (Hint. Consider the function

$$t \longmapsto f(t, \sqrt{1-t^2}) - f(-t, -\sqrt{1-t^2})$$

defined on $[-1, 1]$.)

3.H. Let f be a function on $\mathbb{R}^2$.

We call f *separately continuous* if for every $a \in \mathbb{R}$ the functions $x \to f(x, a)$ and $y \to f(a, y)$ are continuous.

(i) Prove: If f is continuous, then f is separately continuous.

(ii) Consider the function $f_0 : \mathbb{R}^2 \to \mathbb{R}$:

$$f_0(x, y) := \frac{xy}{x^2 + y^2} \quad \text{if } (x, y) \neq (0, 0),$$

$$f_0(0, 0) := 0.$$

Show that f_0 is separately continuous but *not* continuous.

4

CHAPTER

Curves in the Plane

So far, our arguments mainly dealt with analysis. In this chapter, we make a bigger step in the direction of topology. In fact, we will prove a famous topological theorem (Brouwer's Fixed Point Theorem) and formulate a second one (Jordan's Closed Curve Theorem). A complete proof of the latter will be given in Chapter 16.

Curves

4.1

Suppose a ladybug walks on the patio from some point A to point B. As a thought experiment, we use a marker and at each time t (starting at $t = 0$ in A) mark the insect's position with the number t:

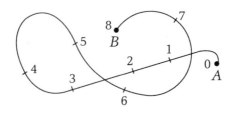

When the ladybug arrives at B we have traced her path. The same effect, tracing a path, can actually be achieved by following the slimy trail of a slug:

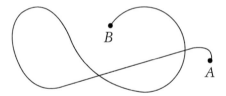

There is an important difference, though. By looking at the trail of the slug, we probably cannot tell if it stopped at some point on the way and did not move for half an hour; nor can we tell if it retraced its path back and forth several times.

It is for those reasons that we are more interested in the map

$$t \mapsto \quad \text{position of the ladybug at time } t \qquad (*)$$

than in the path without the time evolution. To give an explicit example,

$$t \mapsto (\cos 2\pi t, \sin 2\pi t) \qquad (t \in [0, 1])$$

could be the description of the slug's voyage. The trail that is left is the unit circle

$$\{x \in \mathbb{R}^2 : \|x\| = 1\}$$

indistinguishable from the trail that would be left if the walk were described by

$$t \mapsto (\cos 4\pi t, \sin 4\pi t) \qquad (t \in [0, 1])$$

or

$$t \mapsto (\sin 2\pi t, \cos 2\pi t) \qquad (t \in [0, 1]).$$

We will define a "curve" to be a description of a walk in the sense of $(*)$. In mathematical language, a curve is a map defined on a closed bounded interval I with values in $\mathbb{R}^2$. This is not our formal definition yet, because we require the map $(*)$ to be continuous. The following definition is just what you expect.

4.2
Let $D \subset \mathbb{R}$ and $f : D \to \mathbb{R}^2$. We call f *continuous* at a point a of D if

$$\text{for every } \varepsilon > 0, \text{ there exists a } \delta > 0 \text{ such that}$$

$$x \in D, \ |x-a| < \delta \quad \Longrightarrow \quad \|f(x)-f(a)\| < \varepsilon.$$

f is said to be *continuous* if it is continuous at every point of D.

Precisely as in Theorems 1.12 and 3.20 we have equivalence of continuity and sequential continuity:

Theorem 4.3
Let $D \subset \mathbb{R}$, $f:D \to \mathbb{R}^2$, and $a \in D$. Then the conditions (α) and (β) are equivalent.

(α) *f is continuous at a.*
(β) *For every sequence $x_1, x_2, \ldots$ in D with $x_n \to a$, we have $f(x_n) \to f(a)$.*

Proof
Left to the reader.   ∎

4.4
Let $D \subset \mathbb{R}$ and $f : D \to \mathbb{R}^2$. For $x \in D$, let $f_1(x)$ and $f_2(x)$ be the coordinates of the vector $f(x)$:

$$f(x) = \big(f_1(x), f_2(x)\big) \qquad (x \in D).$$

Then f_1 and f_2 are functions on D with values in $\mathbb{R}$, the *component functions* of f.

If $a \in D$ and if $x_1, x_2, \ldots$ is a sequence in D that converges to a, then (by Theorem 3.16)

$$f(x_n) \to f(a) \iff f_1(x_n) \to f_2(a) \text{ and } f_2(x_n) \to f_2(a).$$

Consequently:

Theorem 4.5
Let $D \subset \mathbb{R}$, $f:D \to \mathbb{R}^2$, and $a \in D$. Then f is continuous at a if and only if both of its component functions are.

4.6
Now we can define curve officially: A *curve* is a continuous map $I \to \mathbb{R}^2$ where I is a closed bounded interval in $\mathbb{R}$.

If $\gamma : I \to \mathbb{R}^2$ is a curve, then we denote by γ^* the image of I under γ:

$$\gamma^* := \{\gamma(t) : t \in I\}.$$

(γ^* is the trail left by the slug.)

If $\gamma : [a, b] \to \mathbb{R}^2$ is a curve, then $\gamma(a)$ and $\gamma(b)$ are called the *initial point* and the *end point* of γ, respectively. (Occasionally, we will speak of the initial and the end point of γ^*. In spite of our discussion at the start of this chapter, much of our intuition about curves γ will be guided by what γ^* looks like.)

A curve $\gamma : [a, b] \to \mathbb{R}^2$ is *closed* if $\gamma(a) = \gamma(b)$.

Some examples of curves:

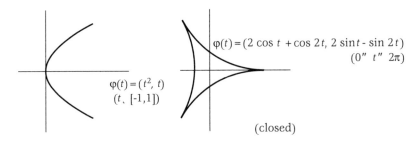

$\varphi(t) = (t^2, t)$
$(t. [-1,1])$

$\varphi(t) = (2\cos t + \cos 2t,\ 2\sin t - \sin 2t)$
$(0'' \ t'' \ 2\pi)$

(closed)

A curve may not be given by a simple analytic formula. The circum-ference of the square $[0, 1] \times [0, 1]$ is the image of γ^* of a (closed) curve $\gamma : [0, 4] \to \mathbb{R}^2$ described by

$$\gamma(t) = (t, 0) \qquad \text{if} \quad 0 \le t \le 1,$$
$$\gamma(t) = (1, t-1) \qquad \text{if} \quad 1 \le t \le 2,$$
$$\gamma(t) = (3-t, 1) \qquad \text{if} \quad 2 \le t \le 3,$$
$$\gamma(t) = (0, 4-t) \qquad \text{if} \quad 3 \le t \le 4.$$

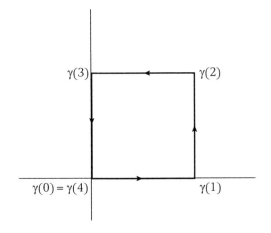

The formula

$$\gamma(t) = (2, 2) \qquad (0 \le t \le 1)$$

describes a rather trivial kind of curve for which γ^* contains only one point: a *point curve*.

It is no surprise that each of the pictures below is a γ^*:

There is, however, one very disconcerting fact about curves that destroys much of this intuition: the following picture is a valid γ^*.

This fact is important enough to discuss it in detail.

Theorem 4.7 (G. Peano)
There exists a curve $\gamma:[0,1] \to \mathbb{R}^2$ such that $\gamma^* = [0,1] \times [0,1]$.

Proof
The construction of the curve γ takes place in stages. At each stage we simultaneously subdivide the interval $[0, 1]$ and the square $[0, 1] \times [0, 1]$.

In stage 1, $[0, 1]$ is divided into four parts, $[0, \frac{1}{4}]$, $[\frac{1}{4}, \frac{1}{2}]$, $[\frac{1}{2}, \frac{3}{4}]$, and $[\frac{3}{4}, 1]$, named I_0, I_1, I_2, and I_3, respectively. Also, $[0, 1] \times [0, 1]$ is divided into four equal parts as below.

S_3	S_2
S_0	S_1

In stage 2, each I_i and each S_i is again divided into four parts, $I_{i0}, \ldots, I_{i3}$ and $S_{i0}, \ldots, S_{i3}$. For instance, the subdivision of I_2 is as follows:

I_{20}	I_{21}	I_{22}	I_{23}
$\dfrac{1}{2}$	$\dfrac{9}{16}$	$\dfrac{10}{16}$	$\dfrac{11}{16}$	$\dfrac{3}{4}$

The subdivision of the square in this stage is

S_{33}	S_{30}	S_{23}	S_{22}
S_{32}	S_{31}	S_{20}	S_{21}
S_{01}	S_{02}	S_{13}	S_{12}
S_{00}	S_{03}	S_{10}	S_{11}

In stage 2, we numbered the subintervals of $[0, 1]$ from left to right. For the subsquares we have to explain our procedure. Note that any two intervals that are consecutive in

$$I_{00}, I_{01}, I_{02}, I_{03}, I_{10}, \ldots, I_{33}$$

have a common end point. We wish to number the subsquares such that two squares consecutive in

$$S_{00}, S_{01}, S_{02}, \ldots, S_{33}$$

border on each other. Second (as noted above), we want $S_{ij} \subset S_i$ for all i and j, precisely like $I_{ij} \subset I_i$. If we add one more wish, namely that S_{00} and S_{33} lie in the lower and the upper left corners, respectively, the numbering is forced on us. The following picture illustrates how the chain of subsquares $S_{00}, S_{01}, \ldots, S_{33}$ is fit into the big square:

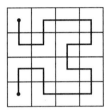

Continuing in this fashion, in stage 3 we make intervals $I_{ij0}, \ldots, I_{ij3}$ and squares $S_{ij0}, \ldots, S_{ij3}$. The chain of little squares now looks like

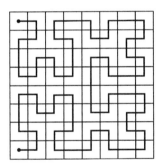

In stage n, we have 4^n subintervals of $[0, 1]$ and 4^n subsquares of $[0, 1] \times [0, 1]$. Each subinterval and each subsquare are numbered by a string of n digits from the set $\{0, 1, 2, 3\}$. If two intervals are neighbors, then so are the identically named subsquares.

Now we construct the curve $\gamma : [0, 1] \to [0, 1] \times [0, 1]$. The idea of the construction is simply this: If a point x of $[0, 1]$ lies in the interval $I_{i_1 \ldots i_n}$, we want $\gamma(x)$ to lie in the corresponding square $S_{i_1 \ldots i_n}$.

Take $t \in [0, 1]$. Choose $i_1, i_2, \ldots$ in $\{0, 1, 2, 3\}$ such that $t \in I_{i_1 \ldots i_n}$ for every n. Then

$$S_{i_1} \supset S_{i_1 i_2} \supset S_{i_1 i_2 i_3} \supset \cdots \quad .$$

It follows from Cantor's Theorem (2.19) that these squares have a common point, x, say. It would be rash to define $\gamma(t) := x$ because the sequence $i_1, i_2, \ldots$ may not be unique. (Indeed, for $t = \frac{1}{2}$ we could have chosen $i_1 = 1$, $i_2 = i_3 = \cdots = 3$ or $i_1 = 2$, $i_2 = i_3 = \cdots = 0$.) However, suppose that besides $i_1, i_2, \ldots$ and x as above, we have another sequence $j_1, j_2, \ldots$ in $\{0, 1, 2, 3\}$ and a $y \in \mathbb{R}^2$ such that $t \in I_{j_1 \ldots j_n}$ and $y \in S_{j_1 \ldots j_n}$ for all n. Then, for each n, the intervals $I_{i_1 \ldots i_n}$ and $I_{j_1 \ldots j_n}$ are either equal or adjacent; then so are the squares $S_{i_1 \ldots i_n}$ and $S_{j_1 \ldots j_n}$; then $\|x-y\| \le 2^{-n}\sqrt{5}$. It follows that $x = y$.

We see that for every $t \in [0, 1]$, there is a unique $\gamma(t) \in [0, 1] \times [0, 1]$ such that

$$t \in I_{i_1 \ldots i_n} \quad \Longrightarrow \quad \gamma(t) \in S_{i_1 \ldots i_n}. \qquad (\star)$$

This defines our map γ.

To see that γ is continuous, take $s, t \in [0, 1]$, $s \ne t$, $|s-t| \le \frac{1}{4}$. There is an $n \in \mathbb{N}$ with $4^{-(n+1)} < |s-t| \le 4^{-n}$. Then s and t lie in the same or in adjacent intervals of stage n, so $\gamma(s)$ and $\gamma(t)$ lie in the same or adjacent squares of stage n. Then

$$\|\gamma(s)-\gamma(t)\| \le 2^{-n}\sqrt{5} = 2\sqrt{5}\, 2^{-(n+1)} \le 2\sqrt{5}\,|s-t|.$$

Thus, γ is continuous.

To see that γ maps the interval *onto* the square, take $x \in [0, 1] \times [0, 1]$. There exist $i_1, i_2, \ldots$ in $\{0, 1, 2, 3\}$ with $x \in S_{i_1 \ldots i_n}$ for all n. Then

$$I_{i_1} \supset I_{i_1 i_2} \supset I_{i_1 i_2 i_3} \supset \cdots \quad .$$

By Cantor's Theorem, these intervals have a common point t. It follows from $(\star)$ that $\gamma(t) = x$. ∎

4.8

There are many variations on this theme. We may for instance subdivide triangles instead of squares:

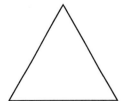 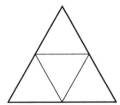 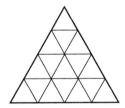

This construction leads to a curve that fills a triangle.

A curve γ for which γ^* contains an entire disk is called a *Peano curve*; the specific one we made in the proof of Theorem 4.7 is the *Hilbert curve*. When dealing with curves, one often has to add extra restrictions in order to exclude Peano curves and remain close to the intuition. In this direction, it is helpful to observe that the Hilbert curve keeps turning back on itself and, thus, is not injective. In fact, the reader can calculate that the elements $\frac{1}{6}$, $\frac{1}{2}$, and $\frac{5}{6}$ of $[0, 1]$ are mapped onto the same point of the square, namely its center. That is not accidental: Peano curves cannot be injective.

4.9

A "Jordan curve" is a curve that is injective except possibly at the end points. More exactly, a curve $\gamma : [a, b] \to \mathbb{R}^2$ is called a *Jordan curve* if

$$\gamma(s) = \gamma(t) \implies s = t \quad \text{or} \quad \begin{matrix} s = a \\ t = b \end{matrix} \quad \text{or} \quad \begin{matrix} s = b \\ t = a \end{matrix}.$$

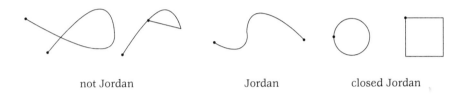

not Jordan   Jordan   closed Jordan

In 4.10 - 4.12 we have a digression, discussing somewhat loosely a few topics of interest which deal with curves.

4.10

An immediate observation about the circle is that it divides the plane into two regions to each of which it is the boundary. This fact is intuitively so clear that it may be a bit of surprise that the same fact is much harder to see for some other closed Jordan curves. Consider this picture. Is P inside the curve or outside?

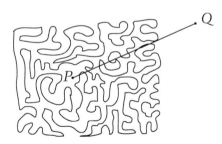

Of course, the point Q is outside the curve. If we slowly shift Q into the direction of P, as indicated in the picture, we notice the following. As soon as we cross the curve we are inside, and as soon as we cross it a second time we are outside again. Continuing in this fashion we discover that P lies outside the curve.

More generally, suppose we are working with a closed Jordan curve. Given a point in the plane, not on the curve, we draw a half-line starting at that point, in such a direction that nowhere the half-line is tangent to the curve. We count the number of intersections with the curve. If the number is odd, we color the point red, otherwise we color it black. (The points on the curve are left white.) The set of black points then is the outside of the curve, the set of red points its inside.

For many closed Jordan curves this procedure works fine to distinguish between inside and outside. This does not constitute a proof that it is always applicable though.

It turns out that the following theorem is difficult to prove. In fact, at this stage we cannot even present a good formulation.

Theorem 4.11 (Jordan's Closed Curve Theorem; preliminary form)
Every closed Jordan curve divides the plane into two nonempty disjoint regions.

At the end of this chapter we will make the statement more precise. We postpone our proof until Chapter 16; it requires machinery that we have yet to develop.

4.12
Another example of the kind of topic that we are going to investigate is provided by the game "Twixt." The game was sold by the Avalon Hill Game Company in the seventies.

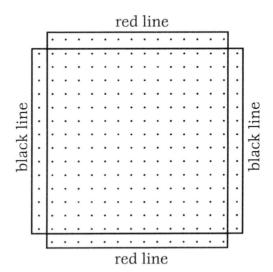

The game is played with red and black pegs and a board provided with holes. There are two players, Red and Black. Players move alternately by placing one peg of their own color in an empty hole that is not on the opponent's color line. Whenever a player has two pegs in a six-hole rectangle as below, he is allowed to link them and thus form

a barrier for his opponent. To win, a player must connect the lines of his color by an uninterrupted chain of linked pegs. Obviously, once Black has completed such a chain, he has a total barrier for Red. Putting it differently, a chain for Black and a chain for Red always intersect.

Or do they? How would you go about proving it?

A red chain from one red line to the other may be described as a β^*, where $\beta : [0, 1] \to [0, 1] \times [0, 1]$ is a curve, beginning somewhere in $[0, 1] \times \{0\}$ and ending in $[0, 1] \times \{1\}$. Similarly, a complete black chain is a γ^*, where $\gamma : [0, 1] \to [0, 1] \times [0, 1]$ is a curve beginning in $\{0\} \times [0, 1]$ and ending in $\{1\} \times [0, 1]$. The previous discussion suggests that β^* and γ^* necessarily intersect. But can we prove that? In Theorem 4.26 we will present a proof. Curiously, it involves complex numbers and continuous maps $\Delta \to \Delta$ where Δ is the "closed unit disk":

$$\Delta := \{x \in \mathbb{R}^2 : \|x\| \leq 1\}.$$

The complex numbers appear only in a supporting role; they can be avoided. The maps $\Delta \to \Delta$, however, are vital.

Homeomorphic Sets

We are compelled for a short while to return to our general theory of continuity. We have considered $\mathbb{R}$-valued functions defined on subsets of $\mathbb{R}$ and of $\mathbb{R}^2$. We have also worked with continuous $\mathbb{R}^2$-valued maps defined on intervals. Now we are going to look at $\mathbb{R}^2$-valued maps defined on subsets of $\mathbb{R}^2$. For the time being, there are no surprises. Indeed, 4.13 and 4.14 are as inevitable as Fate in a Greek tragedy.

4.13
Let $D \subset \mathbb{R}^2$ and $f : D \to \mathbb{R}^2$. We call f *continuous at* a point a of D if

for every $\varepsilon > 0$, there exists a $\delta > 0$ such that

$$x \in D, \ \|x-a\| < \delta \ \implies \ \|f(x)-f(a)\| < \varepsilon.$$

f is said to be *continuous* if it is continuous at every point of D.

4.14
Again (see Theorems 1.12, 3.20, and 4.3), continuity is the same as sequential continuity. This time we do not even bother to formulate a theorem.

As in 4.4, every $f : D \rightarrow \mathbb{R}^2$ gives rise to two $\mathbb{R}$-valued functions f_1 and f_2 on D, its component functions. f is continuous if and only if f_1 and f_2 are.

With the aid of sequential continuity (or directly from the definition) one easily proves:

Theorem 4.15
If $D,E \subset \mathbb{R}^2$, if $f : D \rightarrow \mathbb{R}^2$ and $g : E \rightarrow \mathbb{R}^2$ are continuous, and if $f(D) \subset E$, then $g \circ f$ is continuous.

Examples 4.16

(i) For every c in $\mathbb{R}^2$, we have a translation

$$T_c : x \longmapsto x+c \qquad (x \in \mathbb{R}^2).$$

If $x, y \in \mathbb{R}^2$, then $\|T_c(x) - T_c(y)\| = \|x - y\|$. It follows that T_c is continuous. (For given $\varepsilon > 0$, simply take $\delta = \varepsilon$.) Observe that T_c is bijective; its inverse is the translation T_{-c}.

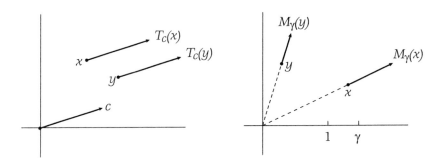

(ii) For every real number γ there is a multiplication

$$M_\gamma : x \longmapsto \gamma x \qquad (x \in \mathbb{R}^2).$$

For $x, y \in \mathbb{R}^2$, we have $\|M_\gamma(x)-M_\gamma(y)\| = \|\gamma x-\gamma y\| = |\gamma| \, \|x-y\|$. It follows that M_γ is continuous. (For given ε, take δ such that $|\gamma|\delta < \varepsilon$; if now $\|x-a\| < \delta$, then $\|M_\gamma(x)-M_\gamma(a)\| < \varepsilon$.) If $\gamma \neq 0$, then M_γ is a bijection whose inverse is the multiplication $M_{1/\gamma}$.

(iii) Take α in $\mathbb{R}$. The formula

$$R_\alpha(x_1, x_2) := (x_1 \cos \alpha - x_2 \sin \alpha \, , \; x_1 \sin \alpha + x_2 \cos \alpha)$$

describes a counterclockwise rotation over the angle α:

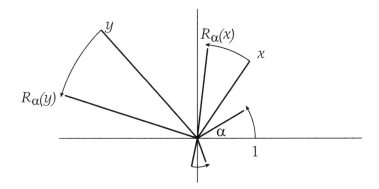

Rotations preserve distances: $\|R_\alpha x - R_\alpha y\| = \|x - y\|$. Hence, R_α is continuous ($\delta = \varepsilon$, again). R_α is bijective; its inverse is $R_{-\alpha}$.

(iv) If φ and ψ are continuous functions $\mathbb{R} \to \mathbb{R}$, then the map $\mathbb{R}^2 \to \mathbb{R}^2$ defined by

$$(x_1, x_2) \longmapsto \big(\varphi(x_1), \psi(x_2)\big) \qquad \big((x_1, x_2) \in \mathbb{R}^2\big)$$

is sequentially continuous, hence continuous. Let us consider a special case that will be of use later. Define $\varphi : \mathbb{R} \to \mathbb{R}$ by

$$\varphi(t) := t^2 \quad \text{if} \quad t \in [0, \infty),$$
$$\varphi(t) := -t^2 \quad \text{if} \quad t \in (-\infty, 0).$$

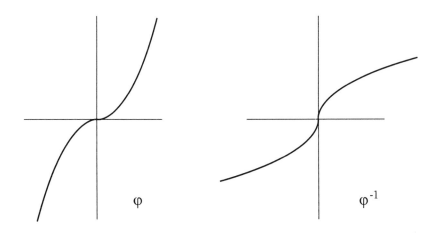

φ is easily seen to be continuous. It is bijective; its inverse map is given by

$$\varphi^{-1}(s) = \sqrt{s} \quad \text{if} \quad s \in [0, \infty),$$
$$\varphi^{-1}(s) = -\sqrt{-s} \quad \text{if} \quad s \in (-\infty, 0),$$

and is also continuous. (Do not take our word for all that.)

As a consequence, the formula

$$F(x_1, x_2) := \big(\varphi(x_1), \varphi(x_2)\big) \qquad \big((x_1, x_2) \in \mathbb{R}^2\big)$$

determines a continuous bijection $F : \mathbb{R}^2 \to \mathbb{R}^2$ with a continuous inverse given by

$$F^{-1}(x_1, x_2) = \big(\varphi^{-1}(x_1), \varphi^{-1}(x_2)\big) \qquad \big((x_1, x_2) \in \mathbb{R}^2\big).$$

4.17

We can now briefly return to our considerations of 1.1 and show that a triangle and a square are "topologically equivalent." First, a definition.

Let D_1 and D_2 be subsets of $\mathbb{R}^2$. A bijective map $f : D_1 \to D_2$ that is continuous and has a continuous inverse $f^{-1} : D_2 \to D_1$ is called a *homeomorphism* of D_1 onto D_2. If, for given D_1 and D_2, such an f exists, we say that D_1 is *homeomorphic* to D_2:

$$D_1 \sim D_2.$$

Observe that then f^{-1} is a homeomorphism of D_2 onto D_1; hence,

$$\text{if } D_1 \sim D_2, \text{ then } D_2 \sim D_1.$$

(We then also say that D_1 and D_2 are homeomorphic.)

If D_1, D_2, and D_3 are subsets of $\mathbb{R}^2$ and if $f : D_1 \to D_2$ and $g : D_2 \to D_3$ are homeomorphisms, then their composition $g \circ f$ is a homeomorphism $D_1 \to D_3$. (Why?) Thus,

$$\text{if } D_1 \sim D_2 \text{ and } D_2 \sim D_3, \text{ then } D_1 \sim D_3.$$

4.18

In Examples 4.16, we have obtained a number of homeomorphisms $\mathbb{R}^2 \to \mathbb{R}^2$. From them, we can construct homeomorphisms between proper subsets of $\mathbb{R}^2$ as follows.

Let f be a homeomorphism of $\mathbb{R}^2$ onto $\mathbb{R}^2$. Take a subset D of $\mathbb{R}^2$. The restriction of f to D [see 1.14(v)] is a bijective map of D onto $f(D)$ whose inverse is the restriction of f^{-1} to $f(D)$. As f and f^{-1} are (sequentially) continuous, so are their restrictions. (Think about this.) Hence, $f|_D$ is a homeomorphism, and

$$D \sim f(D).$$

Example 4.19

We are interested in the *closed unit disk*

$$\Delta := \{x \in \mathbb{R}^2 : \|x\| \le 1\},$$

the *closed unit square*

$$S := [-1, 1] \times [-1, 1],$$

the diamond

$$D := \{(x_1, x_2) \in \mathbb{R}^2 : |x_1| + |x_2| \leq 1\}$$

and the triangle

$$T := \{(x_1, x_2) \in \mathbb{R}^2 : x_1 \geq 0, \ x_2 \geq 0, \ x_1 + x_2 \leq 1\}.$$

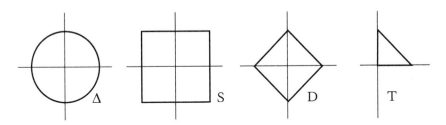

The rotation $R_{\pi/4}$ followed by the multiplication $M_{\sqrt{2}}$ maps the diamond onto the square, i.e.,

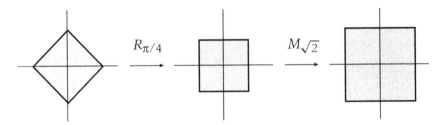

$(M_{\sqrt{2}} \circ R_{\pi/4})(D) = S$. Therefore, the diamond and the square are homeomorphic:

$$D \sim S.$$

Not exactly startling. More interesting is what the map F of 4.16(iv) does to the disk. Indeed, for $x = (x_1, x_2) \in \mathbb{R}^2$, we have

$$x \in \Delta \iff x_1^2 + x_2^2 \leq 1$$
$$\iff |\varphi(x_1)| + |\varphi(x_2)| \leq 1 \iff F(x) \in D.$$

Thus, $F(\Delta) = D$. Then the disk is homeomorphic to the diamond and, therefore, also to the square:

$$\Delta \sim D, \qquad \Delta \sim S.$$

Next, let $\mathbb{R}_+^2$ be the first quadrant,

$$\mathbb{R}_+^2 := \{(x_1, x_2) \in \mathbb{R}^2 : x_1 \geq 0, \ x_2 \geq 0\},$$

and set $\Delta_+ := \Delta \cap \mathbb{R}_+^2$:

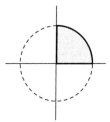

It is easy to see that $F(\mathbb{R}^2_+) = \mathbb{R}^2_+$, so that $F(\Delta_+) = F(\Delta) \cap F(\mathbb{R}^2_+) = D \cap \mathbb{R}^2_+ = T$. We obtain

$$\Delta_+ \sim T.$$

On the other hand, if A is the "angle"

$$A := \{(x_1, x_2) \in \mathbb{R}^2 : x_2 \geq |x_1|\},$$

then

$$x \in A \iff F(x) \in A$$

(check this) and, therefore, $F(\Delta \cap A) = D \cap A$. We have a chain of homeomorphisms

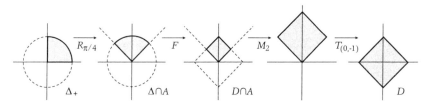

showing that Δ_+ and D are homeomorphic.

Thus, *the sets D, S, Δ, Δ_+, and T are pairwise homeomorphic.*

Brouwer's Theorem

So far it was relatively smooth sailing. We will have to work considerably harder to prove the following powerful statement.

Theorem 4.20 (Brouwer's Fixed Point Theorem)

Let Δ be the closed unit disk. Let $f : \Delta \rightarrow \mathbb{R}^2$ be continuous and such that $f(\Delta) \subset \Delta$. Then there exists an element x in Δ for which $f(x) = x$ (a fixed point for f).

4.21 Intuitive proof

The exact proof, although only moderately complicated, does not show very well what is going on. Therefore, we first do some reconnaissance, without bothering much about exactness. The formal proof follows in 4.24.

Suppose we have a continuous map $f : \Delta \rightarrow \Delta$ where Δ is the closed unit disk, and *suppose $f(x) \neq x$ for all x in Δ.*

For $t \in [0, 1]$, the point

$$\gamma(t) := (\cos 2\pi t, \sin 2\pi t)$$

lies on the unit circle. For each t, we draw an arrow from $\gamma(t)$ to $f(\gamma(t))$; this arrow

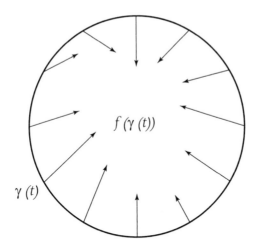

always points inward. If we let t increase from 0 to 1, we see $\gamma(t)$ describe the circle in the counterclockwise direction and we see the arrow make a complete turn. (It may not move regularly: It may lengthen and shorten and it may shift back now and then, but the net result will be one full turn.) Then the point

$$f(\gamma(t)) - \gamma(t)$$

describes a path in the plane that encloses the origin. Putting it differently, for each t

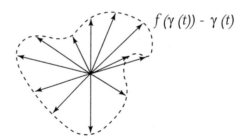

the vector $f(\gamma(t)) - \gamma(t)$ makes a sharp angle with $-\gamma(t)$; as t moves from 0 to 1, $-\gamma(t)$ turns once around the origin; then so does $f(\gamma(t)) - \gamma(t)$.

Next, for $r \in [0, 1]$, we consider the curve

$$\beta_r(t) := f(r\gamma(t)) - r\gamma(t).$$

β_1 is the curve sketched in $(*)$; it encloses the origin. If we let r shrink from 1 to 0, the curve β_r changes gradually *without ever passing through the origin*. Consequently, *every β_r goes around the origin*. However, β_0 is the point curve

$$t \mapsto f(0) \qquad (0 \le t \le 1)$$

and does *not* go around the origin. *Contradiction.* [Or, if you prefer, for very small r, β_r is a curve in the immediate neighborhood of the point $f(0)$ that is not the origin, so β_r cannot enclose the origin.]

4.22

To make an exact proof out of these considerations we will, for any closed curve that does not pass through the origin, formally define its "winding number" which counts how often the curve goes around the origin in the counterclockwise direction. With β_r as above, we will then show that the winding numbers of β_1 and β_0 are 1 and 0, respectively, and that the winding number of β_r depends continuously on r. As winding numbers are integers, we have a contradiction with the Intermediate Value Theorem.

In order to show what (intuitively) the winding number is, we sketch a few curves indicating with each one its winding number n.

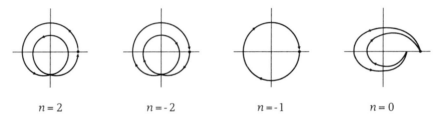

$n = 2$ $n = -2$ $n = -1$ $n = 0$

For the exact definition we need the "argument."

A number ϑ is called an *argument* of the point $x \in \mathbb{R}^2$, $x \neq 0$, if

$$x = \|x\|(\cos \vartheta, \sin \vartheta),$$

i.e. if $\|x\|$ and ϑ form a set of polar coordinates for x. Every nonzero point of $\mathbb{R}^2$ has infinitely many arguments; any two of these differ by a multiple of 2π. (Our use of the term "argument" may not be precisely the one you have seen before.)

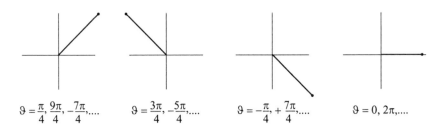

$$\vartheta = \frac{\pi}{4}, \frac{9\pi}{4}, -\frac{7\pi}{4}, ... \qquad \vartheta = \frac{3\pi}{4}, -\frac{5\pi}{4}, ... \qquad \vartheta = -\frac{\pi}{4}, +\frac{7\pi}{4}, ... \qquad \vartheta = 0, 2\pi, ...$$

Let $\gamma : [a, b] \to \mathbb{R}^2$ be a closed curve, $\gamma(t) \neq 0$ for every t. As t increases from a to b, $\gamma(t)$ moves through the plane, starting at $\gamma(a)$ and returning to the same point. For every t, one can choose an argument $\vartheta(t)$ of $\gamma(t)$. Is it possible to do so in such a way that $\vartheta(t)$ depends continuously on t? Intuitively, you would answer affirmatively. You would be right, but a proof is not at all obvious.

Suppose we have such a continuous selection of arguments; i.e., suppose we have a continuous function $\vartheta : [a, b] \to \mathbb{R}$ such that for every t,

$$\gamma(t) = \|\gamma(t)\|\big(\cos \vartheta(t), \sin \vartheta(t)\big).$$

As $\gamma(a) = \gamma(b)$, we see that $\cos \vartheta(a) = \cos \vartheta(b)$ and $\sin \vartheta(a) = \sin \vartheta(b)$, so that $\vartheta(b) - \vartheta(a)$ is a multiple of 2π:

$$\vartheta(b) - \vartheta(a) = 2\pi n \quad \text{for some} \quad n \in \mathbb{Z}.$$

For an example, consider the following curve:

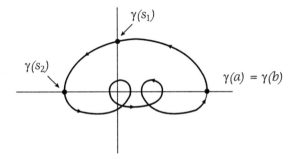

The intersections of γ^* with the coordinate axes are $\gamma(a)$, $\gamma(s_1)$, $\gamma(s_2)$, $\ldots$, $\gamma(b)$, where $a < s_1 < s_2 < \cdots < b$. Starting with $\vartheta(a) = 0$, we see that necessarily $\vartheta(s_1) = \pi/2$, $\vartheta(s_2) = \pi$, $\ldots$, $\vartheta(b) = 4\pi$. We obtain $n = 2$.

Looking at the picture, you will agree that 2 is also what we would want the winding number of this curve to be. That is not accidental. Use the same technique to "calculate" $(2\pi)^{-1}(\vartheta(b) - \vartheta(a))$ for the four curves sketched at the beginning of this section and in each case you will find the hoped-for winding number. Moreover, you will get some idea of why this should be so.

As mathematicians we now *define* the *winding number* of a closed curve $\gamma : [a, b] \rightarrow \mathbb{R}^2$ that avoids the origin to be the integer $(2\pi)^{-1}(\vartheta(b)-\vartheta(a))$ if ϑ is a continuous function and $\vartheta(t)$ is an argument of $\gamma(t)$ for every t. (If we have two such functions, then their difference is a continuous function whose values are multiples of 2π; then this difference is constant and the winding number does not depend on the choice of ϑ.)

It is not at all evident that there always is such a function ϑ. Its existence is, however, an easy consequence of the following lemma.

Lemma 4.23
Let R be the square $[0,1] \times [0,1]$ in $\mathbb{R}^2$. Let g be a continuous map of R into $\mathbb{R}^2$ with $g(x) \neq 0$ for all $x \in R$. Then there is a continuous $\vartheta{:}R \rightarrow \mathbb{R}$ such that

$$g(x) = \|g(x)\|\big(\cos \vartheta(x), \sin \vartheta(x)\big) \qquad (x \in R),$$

i.e. $\vartheta(x)$ is an argument of $g(x)$ for every x.

Proof

(I) Let us first assume $\|g(x)\| = 1$ for every x.

It will be convenient to identify the points of $\mathbb{R}^2$ with the complex numbers via the formula

$$(x_1, x_2) = x_1 + x_2 i \qquad (x_1, x_2 \in \mathbb{R}).$$

In particular, $(r \cos 2\pi t, r \sin 2\pi t) = re^{2\pi it}$ for all r and t.

What we need is a continuous $\vartheta : R \rightarrow \mathbb{R}$ for which

$$g(x) = e^{i\vartheta(x)} \qquad (x \in S).$$

It is easy to verify that

$$u = \exp[i \sin^{-1}(\operatorname{Im} u)] \text{ if } u \in \mathbb{C}, \ |u| = 1, \ \operatorname{Re} u > 0. \qquad (*)$$

As R is sequentially compact [Example 3.26(i)], g is uniformly continuous (Theorem 3.25), so there is an N in $\mathbb{N}$ with

$$y, z \in R, \ |y-z| \leq \frac{\sqrt{2}}{N} \quad \Longrightarrow \quad |g(y)-g(z)| < 1.$$

Therefore, if $x \in R$ and $n \in \{1, 2, \ldots, N\}$, then

$$\left| g\left(\frac{n}{N}x\right) - g\left(\frac{n-1}{N}x\right) \right| < 1,$$

$$\left| \frac{g(\frac{n}{N}x)}{g(\frac{n-1}{N}x)} - 1 \right| < 1,$$

$$\mathrm{Re}\, \frac{g(\frac{n}{N}x)}{g(\frac{n-1}{N}x)} > 0,$$

and by $(*)$,

$$\frac{g(\frac{n}{N}x)}{g(\frac{n-1}{N}x)} = \exp\left[i \sin^{-1}\left(\mathrm{Im}\, \frac{g(\frac{n}{N}x)}{g(\frac{n-1}{N}x)} \right) \right].$$

It follows that if we select any argument α of $g(0)$ and define

$$\vartheta(x) := \alpha + \sum_{n=1}^{N} \sin^{-1}\left(\mathrm{Im}\, \frac{g(\frac{n}{N}x)}{g(\frac{n-1}{N}x)} \right) \qquad (x \in R),$$

then ϑ is continuous and $e^{i\vartheta(x)} = g(x)$ for all x.

(II) If we do not have $\|g(x)\| = 1$ for all x, we simply apply the result of (I) to the map

$$x \longmapsto \frac{g(x)}{\|g(x)\|} \qquad (x \in R). \qquad\blacksquare$$

4.24 Proof

Now we prove Brouwer's Theorem. Let Δ be the closed unit disk, let $f : \Delta \to \Delta$ be continuous and *suppose* $f(x) \neq x$ for all $x \in \Delta$. We derive a contradiction.

Let R be the square $[0, 1] \times [0, 1]$. For $(r, t) \in R$, define $\beta_r(t)$ as earlier; in the language of complex numbers,

$$\beta_r(t) = f(re^{2\pi it}) - re^{2\pi it}.$$

By Lemma 4.23, there exists a continuous function $\vartheta : R \to \mathbb{R}$ such that for all r and t, $\quad \vartheta(r, t)$ is an argument of $\beta_r(t)$. Then for all r,

$$n_r := (2\pi)^{-1}\big(\vartheta(r, 1) - \vartheta(r, 0) \big)$$

is the winding number of β_r. Clearly, $r \mapsto n_r$ is continuous. Its values are integers. Hence, by the Intermediate Value Theorem,

$$n_1 = n_0.$$

For each t, $\quad \vartheta(0, t)$ is an argument of $f(0)$, so $\vartheta(0, t) - \vartheta(0, 0)$ is a multiple of 2π. Again by the Intermediate Value Theorem, $\vartheta(0, 1) - \vartheta(0, 0) = 0$, so

$$n_0 = 0.$$

On the other hand, for all t,

$$|e^{2\pi it} + \beta_1(t)| = |f(e^{2\pi it})| \le 1 = |e^{2\pi it}|,$$

from which it follows that $\beta_1(t)$ and $e^{2\pi it}$ cannot have the same arguments. Thus, $\vartheta(1, t) - 2\pi t$ cannot be a multiple of 2π. The Intermediate Value Theorem is easily seen to imply that $\vartheta(1, 0) - 2\pi < \vartheta(1, t) - 2\pi t < \vartheta(1, 0) + 2\pi$, from which

$$n_1 = 1.$$

Contradiction.   ∎

4.25
Brouwer's Theorem is not typical for the disk. In Exercise 4.19 we have observed the existence of a homeomorphism of Δ onto the closed unit square S, i.e. a continuous bijection $T : \Delta \to S$ whose inverse $T^{-1} : S \to \Delta$ is also continuous. Let us investigate what happens if $f : S \to S$ is a continuous map. Then $T^{-1} \circ f \circ T$ is continuous $\Delta \to \Delta$. Brouwer's Theorem applies to $T^{-1} \circ f \circ T$ and we get a point x in Δ with $T^{-1} \circ f \circ T(x) = x$. Setting $y := T(x)$, we obtain $y \in S$ and $f(y) = y$: Every continuous map $S \to S$ has a fixed point.

When we reread the previous lines we see that S was not so special either. Indeed, the same reasoning applies to every set that is homeomorphic to Δ. Thus, we get a whole bunch of subsets of $\mathbb{R}^2$ that have the property expressed in Brouwer's Theorem.

In 4.12, in connection with the game of "Twixt," we formulated a problem whose solution had to wait for Brouwer's Theorem. Now we can give a complete answer. (We permit ourselves the liberty of considering curves defined on $[-1, 1]$ instead of $[0, 1]$.)

Theorem 4.26
Let β and γ be curves from $[-1,1]$ into a rectangle $[a,b] \times [c,d]$ with $\beta_2(-1) = c$, $\beta_2(1) = d$, $\gamma_1(-1) = a$, and $\gamma_1(1) = b$. Then β^* intersects γ^*. ($\beta_1, \beta_2, \gamma_1, \gamma_2$ are the component functions of β and γ.)

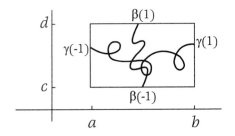

Proof
Assume $\beta^* \cap \gamma^* = \varnothing$. Let S be the closed unit square $[-1, 1] \times [-1, 1]$. For $(s, t) \in S$, define (using the symbol $\vee$ as in 2.4)

$$M(s, t) := |\beta_1(t) - \gamma_1(s)| \vee |\beta_2(t) - \gamma_2(s)|;$$

then $M(s, t) > 0$. Then we can make $f : S \to \mathbb{R}^2$ by

$$f(s, t) := \frac{1}{M(s, t)} \left(\beta_1(t) - \gamma_1(s), \gamma_2(s) - \beta_2(t)\right).$$

f is continuous and maps S into S. By Brouwer's Theorem, applied to S, there exist $\bar{s}, \bar{t} \in [-1, 1]$ with $f(\bar{s}, \bar{t}) = (\bar{s}, \bar{t})$. But f actually maps S into the set $\{(u, v) \in \mathbb{R}^2 : |u| = 1 \text{ or } |v| = 1\}$ (the circumference of S), so $\bar{s} = \pm 1$ or $\bar{t} = \pm 1$. Suppose $\bar{s} = -1$. Then

$$\frac{\beta_1(\bar{t}) - \gamma_1(\bar{s})}{M(\bar{s}, \bar{t})} = -1.$$

But $\beta_1(\bar{t}) - \gamma_1(\bar{s}) = \beta_1(\bar{t}) - \gamma_1(-1) = \beta_1(\bar{t}) - a \geq 0$ and $M(\bar{s}, \bar{t}) > 0$: *contradiction*. The assumptions $\bar{s} = 1$, $\bar{t} = -1$ and $\bar{t} = 1$ lead in similar ways to contradictions. ∎

The theorem shows the relation between the game of Twixt and Brouwer's Theorem. In *The game of Hex and the Brouwer fixed point theorem* (American Mathematical Monthly 86 (1979), 818-827) David Gale shows that a game similar to Twixt has even closer connections with the theorem.

4.27
There are other ways of looking at Theorem 4.26. Suppose, for instance, that γ actually is a Jordan curve.

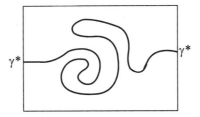

The situation is similar to that of our preamble to the Jordan Curve Theorem. γ^* seems to divide the square into two parts: the points above and the points below γ^*. The previous theorem says (more or less) that a curve connecting a point below γ^* with a point above γ^* has to intersect γ^*.

We have encountered such a phenomenon before in Theorem 2.18. Indeed, let $f : [a, b] \to \mathbb{R}$ be a continuous function. The graph of f is the image of the Jordan curve

$$t \mapsto \big(t, f(t)\big) \qquad (t \in [a, b]).$$

If $f(a) < p < f(b)$, then (a, p) lies above the graph and (b, p) lies below it. These two points are connected by the horizontal line that is the image of the curve $s \mapsto (s, p)$.

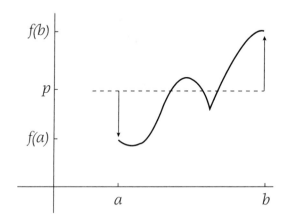

The Intermediate Value Theorem yields a point of intersection.

One could say that a curve such as γ above disconnects the rectangle. Two points in different parts cannot be connected by a β^* without crossing γ^*. These observations lead to the following definition.

4.28
A subset U of $\mathbb{R}^2$ is called *path connected* if for every pair of points $x, y \in U$, there exists a curve $\gamma : [a, b] \to U$ with $\gamma(a) = x$, $\gamma(b) = y$.

It is quite obvious that the inside of a circle is path connected and so is its outside. The upper half-plane $\{(s, t) \in \mathbb{R}^2 : t > 0\}$ is path connected, but $\{(s, t) \in \mathbb{R}^2 : t \neq 0\}$ is not.

We can now be more precise about the formulation of Jordan's Theorem, although its proof will have to wait. (See 16.28.)

Theorem 4.29 (Jordan's Closed Curve Theorem; improved version)
In $\mathbb{R}^2$, the complement of the image of a closed Jordan curve is the union of two disjoint nonempty sets, each of which is path connected.

Extra: L.E.J. Brouwer (1881-1966)

Luitzen Egbertus Jan Brouwer, born in the village Overschie in the Netherlands, became the second Dutch mathematician of worldwide fame.

Before him, the Dutch had produced internationally influential artists and scientists, but after Huygens in the seventeenth century, there had not been a mathematician of his caliber.

Brouwer entered the mathematical arena just before a suitable basis for a study of topology became generally accepted. This new area of mathematics was as yet part of the study of Euclidean spaces, mostly $\mathbb{R}$ or $\mathbb{R}^2$. The name for the arising aggregate of results was *Analysis Situs* (Latin for *analysis of location*). Many of the problems in Analysis Situs had come out of the nineteenth century as "easy to state but hard to work." Brouwer successfully answered several of the open questions with unprecedented precision, conciseness, and clarity.

At this stage in the book we are in a position to understand the problems he attacked, if not their solutions. We describe four of them and note that he solved all four (and others) in a two-year period.

- *The Jordan Closed Curve Theorem.* Oswald Veblen in 1905 was the first to prove the theorem correctly, but in 1912, Brouwer generalized it to higher dimensions (e.g., a subset S of $\mathbb{R}^3$ that is homeomorphic to the sphere $\{x \in \mathbb{R}^3 : \|x\| = 1\}$ divides $\mathbb{R}^3$ into two connected parts each of which has S as its boundary.)
- *Invariance of dimension.* We have seen that there exists a continuous map of $[0, 1]$ onto $[0, 1] \times [0, 1]$ (Peano's curve; Theorem 4.7). Furthermore, Cantor had shown the existence of a bijection $[0, 1] \rightarrow [0, 1] \times [0, 1]$. More generally, for any $n, m \in \mathbb{N}$ one can show that there exist a continuous surjection and also a bijection of $[0, 1]^m$ onto $[0, 1]^n$. Can there exist a continuous bijection if $n \neq m$? Such a continuous bijection would automatically be a homeomorphism (Corollary 13.23). It is easy to see that $[0, 1]$ is not homeomorphic to $[0, 1]^2$, but how about $[0, 1]^2$ and $[0, 1]^3$? In 1910, Brouwer proved that $[0, 1]^n$ and $[0, 1]^m$ are homeomorphic only if $n = m$.
- *Brouwer's Fixed Point Theorem* (1911) has been immensely influential, not only in topology but in many branches of mathematics, e.g., numerical analysis. Again, he generalized the theorem to higher dimensions later.
- *The "Lakes of Wada."* Imagine a map with three countries, each being path connected. Many points will be on the border of two countries and some may be at the border of three; intuition tells us that points of the latter type are rare. Brouwer managed to produce an example of three mutually disjoint path connected regions in the plane having precisely the same boundaries! (A popular description of this example is known as the "Lakes of Wada.")

But Brouwer's main interests were of a philosophical nature. The flurry of topological activities that gave him his fame followed his Ph.D. thesis (1907) "On the Foundations of Mathematics," which was, in part, a plea for a different way of doing mathematics. In it, Brouwer defends the idea that mathematics is an individual activity of the human mind rather than the discovery of outside truths. In his 1908 article "On the Unreliability of Logical Principles" he submits that the existence of a mathematical object has not yet been proved if only its nonexistence has been refuted: A construction is needed. In other words, Brouwer rejects the "argument by contradiction." The consequences of that stance are stupendous. On the one hand, many classical results, such as the Intermediate Value Theorem, become false. On the other hand, in Brouwer's concept of mathematics, "Let x be a real number" means, roughy speaking, "Suppose I have an explicit way to calculate as many decimals of x as I like," which is a very strong assumption. The result is that many Brouwerian theorems are falsehoods in the eyes of other mathematicians. For instance, in Brouwer's theory, *every* function defined on all of $\mathbb{R}$ is continuous!

Ironically (considering Brouwer's ambitions), whereas his topological work has had a vast impact on the mathematical society, his philosophical ideas have by and large been ignored. Few mathematicians have been willing to give up a comfortable way of thinking.

Further Reading

Bell, E.T., *Development of Mathematics*, McGraw-Hill Book Company, 1945.

van Dalen, D., The War of the Frogs and the Mice, or the Crisis of the Mathematische Annalen, *Mathematical Intelligencer* 12 (1990), 17-31.

Exercises

4.A.   (i) Make a continuous map $\Gamma \to \Gamma$ without fixed points.
       (ii) Show that Γ and $[0, 1]$ are not "homeomorphic" in the sense of 4.17. (See Exercise 2.J.)

4.B.   Make a continuous function $(0, 1] \to (0, 1]$ without fixed points.

4.C.   Let Δ be the closed unit disk, let $\Gamma := \{z \in \mathbb{C} : |z| = 1\}$. Let f be a continuous function $\Delta \to \Gamma$. Show that for every $\lambda \in \Gamma$ there is a $u \in \Gamma$ with $f(u) = \lambda u$.

4.D.   Let Δ be the closed unit disk $\{x \in \mathbb{R}^2 : |x| \leq 1\}$ and Γ the unit circle $\{z \in \mathbb{C} : |z| = 1\}$. Let $f : \Delta \to \Gamma$ be continuous. Show the existence of a continuous function $\vartheta : \Delta \to \mathbb{R}$ for which

$$f(x) = e^{i\vartheta(x)} \qquad (x \in \Delta).$$

Postscript. We use this to prove the *Main Theorem of Algebra*: *If $N \in \mathbb{N}$ and $\alpha_1, \ldots \alpha_N \in \mathbb{C}$, then the equation*

$$z^N + \alpha_1 z^{N-1} + \alpha_2 z^{N-2} + \cdots + \alpha_N = 0$$

has a solution.

Let $N, \alpha_1, \ldots, \alpha_N$ be as above. Define $P : \mathbb{C} \to \mathbb{C}$ by

$$P(z) = z^N + \alpha_1 z^{N-1} + \cdots + \alpha_N \quad (z \in \mathbb{C}).$$

First, observe that for $z \neq 0$,

$$|z^{-N}P(z)-1| \leq |\alpha_1||z|^{-1} + |\alpha_2||z|^{-2} + \cdots + |\alpha_N||z|^{-N}.$$

Choose $R > 0$ such that $|\alpha_1|R^{-1} + \cdots + |\alpha_N|R^{-N} < 1$; then

$$|z^{-N}P(z)-1| < 1 \quad \text{if} \quad |z| \geq R. \qquad (*)$$

Suppose P does not take the value 0. It follows from the exercise that there exists a continuous $\vartheta : \Delta \to \mathbb{R}$ with

$$\frac{-P(Rx)}{|P(Rx)|} = e^{i\vartheta(x)} \quad (x \in \Delta).$$

$x \longmapsto e^{i\vartheta(x)/N}$ is a continuous map $\Delta \to \Delta$. By Brouwer's Theorem, there is an $x \in D$ with $x = e^{i\vartheta(x)/N}$. Then $|x| = 1$ and

$$x^N = e^{i\vartheta(x)} = -\frac{P(Rx)}{|P(Rx)|}.$$

Setting $z := Rx$, we obtain $|z| = R$ and

$$\frac{z^N}{|z^N|} = -\frac{P(z)}{|P(z)|},$$

so that $z^{-N}P(z) = -|z^{-N}||P(z)| \in (-\infty, 0)$, contradicting $(*)$.

4.E. (i) Show that there does not exist a continuous $f : \Delta \to \Gamma$ with

$$f(x) = x \quad (x \in \Gamma).$$

(Apply Brouwer's Theorem to $x \mapsto -f(x)$.)

(ii) Deduce from (i): *If* $g:\Delta \to \Delta$ *is continuous and* $g(x) = x$ *for all* $x \in \Gamma$, *then* $\Delta \subset g(\Delta)$.

Hint. *Suppose* $b \in \Delta, b \notin g(\Delta)$. Show, by solving a quadratic equation, that for every $z \in \mathbb{R}^2, z \neq b$, there exists a unique positive number $\lambda(z)$ with $\|b+\lambda(z)(z-b)\| = 1$. Show that $z \mapsto b+\lambda(z)(z-b)$ is a continuous function $j : \mathbb{R}^2\backslash\{b\} \to \Gamma$ with $j(x) = x$ for all $x \in \Gamma$. Now derive a *contradiction* by considering $f := j \circ g$.

4.F. As another application of Brouwer's Theorem we prove the following result, due to O. Perron and L. Frobenius: *Every* 3×3 *matrix whose entries are nonnegative has a nonnegative eigenvalue.*

(i) Let O be the "positive octant of the unit sphere" in $\mathbb{R}^3$:

$$O := \{x \in \mathbb{R}^3 : \|x\| = 1; \text{ all coordinates of } x \text{ are } \geq 0\}.$$

Let T be the triangle described in Example 4.19. Show that the map

$$(s, t) \mapsto (\sqrt{s}, \sqrt{t}, \sqrt{1-s-t}) \quad \left((s, t) \in T\right)$$

is a homeomorphism of T onto O.

(ii) Apparently, O is homeomorphic to T, hence to Δ. It follows that every continuous map $O \to O$ has a fixed point (4.25).

Take a 3×3 matrix A all of whose entries are non-negative. If there is an x in O with $Ax = 0$, then 0 is an eigenvalue of A. Show that, otherwise,

$$x \longmapsto \frac{Ax}{\|Ax\|} \qquad (x \in O)$$

is a continuous map $O \to O$ and that A has a positive eigenvalue.

II

PART

Metric Spaces

5

CHAPTER

Metrics

5.1

We have some experience with the concept of continuity. Loosely speaking, continuity of a function F means

(i) if two elements, x and y, of the domain of F are close together, then the function values $F(x)$ and $F(y)$ will be approximately equal.

So far, x and y have always been real numbers or points in $\mathbb{R}^2$. The same idea, however, crops up in very different circumstances. Consider this statement:

(ii) if two continuous functions on $[0, 5]$ are close together, their integrals will be approximately equal.

For instance, if $|f(t)-g(t)| \leq \frac{1}{100}$ for all $t \in [0, 5]$, then $|\int_0^5 f(t)dt - \int_0^5 g(t)dt| \leq \frac{1}{20}$, and if $|f(t)-g(t)| \leq \frac{1}{1000}$ for all t, then $|\int_0^5 f(t)dt - \int_0^5 g(t)dt| \leq \frac{1}{200}$.

The resemblance between (i) and (ii) is not merely superficial. In statement (ii), one actually regards

$$\int_0^5$$

as a *function* whose domain happens to be itself a set of functions (the continuous functions on $[0, 5]$). Statement (ii) describes continuity of this function

81

$$\int_0^5$$

just as Statement (i) describes continuity of F. Admittedly, it is vague—
but then, so is statement (i). We know how statement (i) can be made
more precise: "For every positive number ε, etc." We can do the same
with statement (ii), provided that we have a way to measure how far two
functions are away from each other.

Now, for this special case, it would not be overly difficult to do so,
but it turns out to be wiser to adopt a more general point of view. The
reason is that in mathematics, one encounters many statements similar
to statement (ii), such as:

(iii) If two parts of the plane closely resemble each other, they will have
approximately the same area.
(iv) If two convergent sequences differ little, they will have approximately equal limits.
 (v) If two convergent series are almost the same, they will have
approximately equal sums.
(vi) Two continuous functions on $[0, 1]$ that are close together will have
almost equal maximal values.

It is possible to define "distances," not only between numbers or points
of $\mathbb{R}^2$ but also between functions, between sequences, and between
subsets of the plane. With the aid of such a definition, statements as
statements (ii)-(vi) can be made more precise. It is the aim of this chapter
to introduce a general concept of "distance" and to see how it may be
applied.

"Distance" has to do, not only with continuity, but also with convergence. For example, if $X_1, X_2, \ldots$ are subsets of $\mathbb{N}$ and if

$$X_1 \subset X_2 \subset X_3 \subset \cdots,$$

then in a very reasonable sense the sequence $(X_n)_{n \in \mathbb{N}}$ may be said to
"converge" to its union $\bigcup_n X_n$: As N gets larger and larger, X_N begins to
look more and more like $\bigcup_n X_n$. Similarly, if

$$X_1 \supset X_2 \supset X_3 \supset \cdots,$$

the sequence $(X_n)_{n \in \mathbb{N}}$ will "converge" to its intersection. We will see that
it is indeed possible to determine a notion of distance between subsets
of $\mathbb{N}$ describing these "convergences" in a natural way.

In daily life the meaning of the word "distance" is far from being
unambiguous.

At first sight it seems quite clear what is meant by the "distance" between two points, A and B, in a room: It is the length of a piece of string
spanned between A and B. But we use a different concept of "distance"
when A is an electrical outlet and B the spot where we want to put a floor

lamp; and a mosquito flying from A to B and having to circumnavigate the furniture uses a third "distance."

By the "distance" between two cities, say Amsterdam and Boston, one usually means the length of a piece of string stretching from A to B over the surface of the Earth, not the length of a needle stuck *through* the Earth.

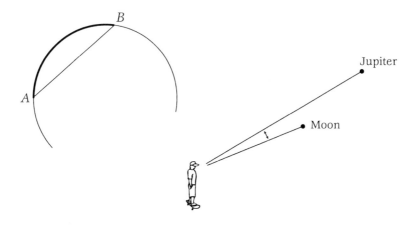

When we look at the night sky and observe Jupiter being close to the Moon, we are not thinking in terms of inches and miles: We mean that a certain angle is small. The distance will be 5 degrees, say, not 5 billion miles (or whatever). Actually, it would be more correct to speak of the distance between two directions than between two celestial objects.

In each of these cases we deal with the distance between two "things" (points in a room, cities, directions) and this distance can be expressed by a number. In other words, we have a set and a real-valued function acting on pairs of elements of it.

The mathematical approach is as follows.

5.2

Let X be a set, d a function on $X \times X$ with values in $\mathbb{R}$. We call d a *distance function* (or a *metric*) if for all $x, y, z \in X$, Axioms M1-M4 are satisfied.

M1	$d(x, y) \geq 0$.
M2	$d(x, y) = 0 \iff x = y$.
M3	$d(x, y) = d(y, x)$.
M4	$d(x, z) \leq d(x, y) + d(y, z)$ ("Triangle Inequality").

These axioms can best be understood in terms of the mosquito example: $d(x, y)$ is the length of the shortest trip from x to y that is feasible for the mosquito. The meanings of Axioms M1 and M2 are obvious. Axiom M4 is intuitively clear, too: The mosquito has a possible path from x to z (via y)

whose length is $d(x, y) + d(y, z)$, so the shortest path from x to z cannot be longer than that. Axiom M3 is valid in the mosquito's world if the length of a trip is measured in feet but not if it is measured in seconds and there is a constant draft in the room.

Asymmetric concepts of distance occur frequently: Just think of one-way traffic or compare the distance from Christmas to New Year's Day and the distance from New Year's Day to Christmas. Axiom M3 says that we are only interested in the distance "between x and y," not "from x to y."

If d is a distance function on a set X (or, rather, on $X \times X$), the couple (X, d) is called a *metric space*. In many cases, however, explicit mention of the metric will be unnecessary and we use phrases like "let X be a metric space."

In 5.3 and 5.5 we present examples of metric spaces. In each case it turns out that the proofs of Axioms M1, M2, and M3 are very simple and we only have to worry (sometimes) about the Triangle Inequality.

Examples 5.3

(i) The basic example is, of course, the *absolute value metric* or *Euclidean metric* d_E on $\mathbb{R}$:

$$d_E(x, y) := |x - y| \qquad (x, y \in \mathbb{R}).$$

(ii) Another metric on $\mathbb{R}$ is given by

$$d(x, y) := |\tan^{-1} x - \tan^{-1} y| \qquad (x, y \in \mathbb{R}).$$

This $d(x, y)$ represents the angle shown in the picture below:

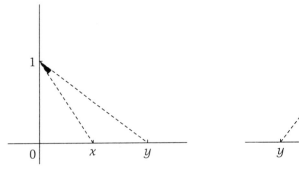 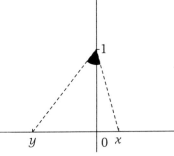

In the sense of this "$\tan^{-1}$ metric," the distance between any two points of $\mathbb{R}$ is less than π, which makes it quite different from the absolute value metric.

(iii) Take $N \in \mathbb{N}$; we consider $\mathbb{R}^N$. (If you feel uneasy with $\mathbb{R}^4$, do not worry; the cases $N = 2$ and $N = 3$ are sufficient for our purposes.) For $x = (x_1, x_2, \ldots, x_N) \in \mathbb{R}^N$, we define its length:

$$\|x\| := (x_1^2 + x_2^2 + \cdots + x_N^2)^{\frac{1}{2}}.$$

In $\mathbb{R}^N$, as in $\mathbb{R}$, we have the *Euclidean metric* d_E:

$$d_E(x, y) := \|x-y\| \qquad (x, y \in \mathbb{R}^N).$$

To show that d_E really is a metric, we have to check the Triangle Inequality. (Statements M1, M2, and M3 are easy to verify.) For that, we use the inner product.

The *inner product* of $x = (x_1, \ldots, x_N) \in \mathbb{R}^N$ and $y = (y_1, \ldots, y_N) \in \mathbb{R}^N$ is the number

$$\langle x, y \rangle := x_1 y_1 + \cdots + x_N y_N.$$

We need *Schwarz' Inequality*:

$$\langle x, y \rangle \leq \|x\| \, \|y\| \qquad (x, y \in \mathbb{R}^N).$$

This inequality is a direct consequence of

$$\|x\|^2 \|y\|^2 - \langle x, y \rangle^2 = \tfrac{1}{2} \sum_{i,j} (x_i y_j - x_j y_i)^2$$

which, in turn, can be calculated by brute force. (Any textbook on linear algebra will give you a more elegant proof, however.) From Schwarz' Inequality, for all $x, y \in \mathbb{R}^N$ one obtains

$$\|x+y\|^2 = \|x\|^2 + \|y\|^2 + 2\langle x, y \rangle$$

$$\leq \|x\|^2 + \|y\|^2 + 2\|x\| \, \|y\| = (\|x\|+\|y\|)^2,$$

from which

$$\|x+y\| \leq \|x\| + \|y\|.$$

From here to the Triangle Inequality is an easy step.

(iv) Let X be the *unit sphere* in $\mathbb{R}^3$:

$$X := \{x \in \mathbb{R}^3 : \|x\| = 1\}.$$

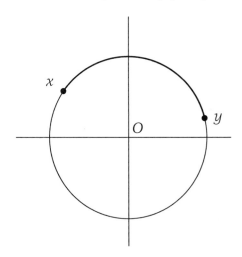

We wish to introduce a metric d on X such that $d(x, y)$ is the length of the "straight" path from x to y that runs over the surface. This length is equal to the angle xOy (measured in radians), i.e., the angle between the directions Ox and Oy.

To arrive at a sound definition, recall that, intuitively, the inner product of x and y [see Example (iii), above] equals $\|x\| \; \|y\| \cos \alpha$, where α is the desired angle. $\|x\|$ and $\|y\|$ being 1, we obtain

$$\cos \alpha = \langle x, y \rangle.$$

Together with the restriction $0 \leq \alpha \leq \pi$, this determines α; apparently, α ought to be $\cos^{-1}\langle x, y \rangle$.

Thus, we are led to the *definition*

$$d(x, y) := \cos^{-1}\langle x, y \rangle \qquad (x, y \in X).$$

Is this d a metric? Again, Axioms M1 and M3 are easy and Axiom M2 is not complicated, but the Triangle Inequality requires work. Let $x, y, z \in X$, $\alpha := d(x, y)$, $\beta := d(y, z)$, and $\gamma := d(x, z)$, i.e.,

$$\left[\begin{array}{l} \alpha, \beta, \gamma \in [0, \pi], \\ \cos \alpha = \langle x, y \rangle, \quad \cos \beta = \langle y, z \rangle, \quad \cos \gamma = \langle x, z \rangle. \end{array} \right.$$

We wish to prove $\gamma \leq \alpha + \beta$. This is trivial in the case $\alpha + \beta \geq \pi$, so we may assume $\alpha + \beta \in [0, \pi]$. Then the desired inequality is equivalent to

$$\cos \gamma \geq \cos(\alpha + \beta).$$

The proof of this formula relies heavily on the properties of the inner product:

$$\left[\begin{array}{l} \langle u+u', v \rangle = \langle u, v \rangle + \langle u', v \rangle \\ \langle u, v+v' \rangle = \langle u, v \rangle + \langle u, v' \rangle \\ \langle \lambda u, v \rangle = \lambda \langle u, v \rangle = \langle u, \lambda v \rangle \end{array} \right.$$

for all $u, u', v, v' \in \mathbb{R}^3$ and $\lambda \in \mathbb{R}$.

Define $x', z' \in \mathbb{R}^3$ by

$$x = (\cos \alpha)y + x', \quad z = (\cos \beta)y - z'.$$

Then

$$\begin{aligned} \langle x', z' \rangle &= \langle x - (\cos \alpha)y, -z + (\cos \beta)y \rangle \\ &= -\langle x, z \rangle + (\cos \alpha)\langle y, z \rangle + (\cos \beta)\langle x, y \rangle - (\cos \alpha)(\cos \beta)\langle y, y \rangle \\ &= -\cos \gamma + \cos \alpha \cos \beta + \cos \alpha \cos \beta - \cos \alpha \cos \beta \\ &= -\cos \gamma + \cos \alpha \cos \beta. \end{aligned}$$

In the same way,

$$\|x'\|^2 = \langle x', x'\rangle = \cdots = 1-(\cos\alpha)^2 = (\sin\alpha)^2$$

so that, since $0 \le \alpha \le \pi$,

$$\|x'\| = \sin\alpha.$$

Similarly,

$$\|z'\| = \sin\beta.$$

But by Schwarz' Inequality, $\langle x', z'\rangle \le \|x'\|\,\|z'\|$. It follows that

$$\cos\gamma \ge \cos\alpha\cos\beta - \sin\alpha\sin\beta = \cos(\alpha+\beta)$$

and we are done.

(v) By $\mathcal{P}(\mathbb{N})$ we denote the set of all subsets of $\mathbb{N}$. Examples of elements of $\mathcal{P}(\mathbb{N})$ are

the set $E = \{2, 4, 6, \ldots\}$ of all even positive integers,
the set $P = \{2, 3, 5, 7, 11, \ldots\}$ of all primes,
the set $S = \{1, 4, 9, \ldots\}$ of all squares,

but also $\varnothing$ and $\mathbb{N}$ itself.

For $A, B \in \mathcal{P}(\mathbb{N})$ we define $d(A, B)$ as follows. If $A = B$, we set, of course, $d(A, B) := 0$. Otherwise, we look for the smallest number that is in one of the sets A and B but not in both; if that is m, then $d(A, B) := m^{-1}$.

Thus, $d(E, P) = \frac{1}{3}$ because

$$1 \notin E \quad\text{and}\quad 1 \notin P,$$

$$2 \in E \quad\text{and}\quad 2 \in P,$$

$$3 \notin E \quad\text{but}\quad 3 \in P.$$

Observe that for $A, B \in \mathcal{P}(\mathbb{N})$ and $m \in \mathbb{N}$ we have

$$d(A, B) < m^{-1} \iff A \cap [1, m] = B \cap [1, m]. \tag{$*$}$$

To show that d is a metric, we only prove the Triangle Inequality. Let $A, B, C \in \mathcal{P}(\mathbb{N})$; we wish to show that

$$d(A, C) \le d(A, B) + d(B, C).$$

Clearly, we may assume $d(A, C) \ne 0$. Then $d(A, C) = m^{-1}$ for some m. By $(*)$, $A \cap [1, m] \ne C \cap [1, m]$. It follows that $A \cap [1, m] \ne B \cap [1, m]$ or $B \cap [1, m] \ne C \cap [1, m]$; but then, $d(A, B) \ge m^{-1}$ or $d(B, C) \ge m^{-1}$. In any case, the Triangle Inequality holds.

(vi) In 5.5 we will consider a number of other metric spaces. At this stage let us mention just one more: For any set X, we have the so-called *trivial metric* d_0 defined by

$$d_0(x, y) := \begin{cases} 1 & \text{if } x \neq y, \\ 0 & \text{if } x = y. \end{cases}$$

5.4

In this Section we collect a few elementary observations. After that, we present some more examples.

Let (X, d) be a metric space.

(i) For all $x, y, z \in X$, we have the Second Triangle Inequality

$$|d(x, z) - d(y, z)| \leq d(x, y).$$

Indeed, by the Triangle Inequality

$$d(x, z) - d(y, z) \leq d(x, y)$$

and, similarly,

$$d(y, z) - d(x, z) \leq d(y, x).$$

But $d(y, x) = d(x, y)$. The Second Triangle Inequality follows.

(ii) For $a \in X$ and $r > 0$, the *ball* on a with *radius* r is the set

$$B_r(a) := B(a; r) := \{x \in X : d(x, a) < r\}.$$

(iii) Let $a, x_1, x_2, \ldots \in X$. We say that the sequence $(x_n)_{n \in \mathbb{N}}$ *converges to* a, or that a is a *limit* of the sequence and write

$$x_n \to a,$$

if

$$\lim_{n \to \infty} d(x_n, a) = 0.$$

This is the case if and only if

for every $\varepsilon > 0$ there is an N

with $n \geq N \implies x_n \in B_\varepsilon(a)$

or, to put it in yet another way,

for every $\varepsilon > 0$, the ball $B_\varepsilon(a)$

contains a tail of the sequence.

The *tails* of the sequence $x_1, x_2, \ldots$ are the sequences $x_N, x_{N+1}, x_{N+2}, \ldots$; see 1.9.

If X is $\mathbb{R}$ or $\mathbb{R}^2$ and if d is the Euclidean metric, then we have just the convergence with which we are familiar.

If the sequence $(x_n)_{n \in \mathbb{N}}$ converges to a, then so does every subsequence. (See Theorem 3.6.)

In any metric space, a sequence can have at most one limit. Indeed, suppose $x_n \to a$ and also $x_n \to b$. For all n,

$$0 \le d(a, b) \le d(a, x_n) + d(x_n, b) = d(x_n, a) + d(x_n, b).$$

Consequently, $d(a, b) = 0$ and $a = b$.

Thus, we may speak of *the* limit of a sequence:

$$\lim_{n \to \infty} x_n.$$

(iv) Notations such as "$B_r(a)$" and "$x_n \to a$" are ambiguous because they make no explicit mention of the metric involved. Occasionally, we write

$$d\text{-}B_r(a), \quad x_n \xrightarrow{d} a, \quad d\text{-}\lim_{n \to \infty} x_n.$$

(v) It is a trivial observation that any subset Y of X inherits an *induced metric* d_Y:

$$d_Y(x, y) := d(x, y) \qquad (x, y \in Y).$$

We have

$$d_Y\text{-}B_r(a) = Y \cap B_r(a) \qquad (a \in Y; \ r > 0)$$

and, for $a, x_1, x_2, \ldots \in Y$,

$$x_n \xrightarrow{d_Y} a \quad \Longleftrightarrow \quad x_n \xrightarrow{d} a.$$

We will not always write the subscript: For $Y \subset X$, we now and then speak of "the metric space (Y, d)."

(vi) As stated in (iv), we sometimes have to mention the metric we use. On the other hand, certain sets carry "natural" metrics. In such cases, mentioning the metric is pointless. Henceforth, when dealing with $\mathbb{R}$ or $\mathbb{R}^N$, it will be tacitly assumed that we are using the Euclidean metric d_E. (Unless, of course, we say otherwise.)

The same applies to subsets of $\mathbb{R}$ and $\mathbb{R}^N$. The expression "the metric space $\mathbb{Q}$" will refer to the set $\mathbb{Q}$ endowed with the metric induced by d_E.

(vii) If (X', d') is a metric space, a map $f : X \to X'$ is said to be an *isometry* if

$$d'\big(f(x), f(y)\big) = d(x, y) \qquad (x, y \in X).$$

(X, d) and (X', d') are *isomorphic (as metric spaces)* if there exists a surjective isometry $f : X \to X'$. Such an f is bijective; its inverse is again an isometry.

Examples 5.5

(i) Let $N \in \mathbb{N}$; we look again at $\mathbb{R}^N$. For $x \in \mathbb{R}^N$, let $(x)_1, (x)_2, \ldots, (x)_N$ be the coordinates of x. (See 3.14.)

Take $a, x_1, x_2, x_3, \ldots \in \mathbb{R}^N$. Just as in 3.14, one sees that

$$\|x_n - a\| \to 0$$

if and only if

$$(x_n)_1 \to (a)_1, \ (x_n)_2 \to (a)_2, \ldots, (x_n)_N \to (a)_N :$$

In $\mathbb{R}^N$, Euclidean convergence simply is coordinatewise convergence.

(ii) Let d be the $\tan^{-1}$ metric in $\mathbb{R}$ [5.3(ii)].

The balls $B_{10}(0)$, $B_9(0)$, and $B_9(3)$ are each equal to the entire space $\mathbb{R}$. Thus, a ball may have many centers and many radii!! (See Exercise 5.B for a more interesting example.)

For $a, x_1, x_2, \ldots \in \mathbb{R}$, we have

$$d(x_n, a) \to 0 \iff \tan^{-1} x_n \to \tan^{-1} a$$

$$\iff x_n \to a$$

$$\iff d_E(x_n, a) \to 0.$$

Thus, two distinct metrics may determine the same convergence.

But if d_0 is the trivial metric on $\mathbb{R}$ [5.3(vi)], then the sequence $(x_n)_{n \in \mathbb{N}}$ d_0-converges to a if and only if there is an N with $x_N = x_{N+1} = x_{N+2} = \cdots = a$. (In the definition of convergence, take $\varepsilon = \frac{1}{2}$.) Apparently, d_0-convergence differs drastically from Euclidean convergence.

(iii) Consider the metric d on $\mathcal{P}(\mathbb{N})$ we have defined in 5.3(v).

For $X, Y \in \mathcal{P}(\mathbb{N})$, $d(X, Y) = d(\mathbb{N} \backslash X, \mathbb{N} \backslash Y)$. Thus, the complementation map $X \mapsto \mathbb{N} \backslash X$ is an isometry $\mathcal{P}(\mathbb{N}) \to \mathcal{P}(\mathbb{N})$.

If $X_1 \subset X_2 \subset X_3 \subset \cdots \subset \mathbb{N}$ and X is the union of the sets X_n, then $X_n \to X$ in the sense of d. (See 5.1.) Indeed, take $\varepsilon > 0$ and $m \in \mathbb{N}$ such that $m^{-1} < \varepsilon$. We see that for large n,

$$X_n \cap [1, m] = X \cap [1, m]$$

so that $d(X_n, X) < m^{-1} < \varepsilon$.

(iv) In spite of the remark we made in 5.4(vi), sometimes one has use for metrics on $\mathbb{R}^N$ that are distinct from the Euclidean one. For instance, on $\mathbb{R}^2$ we have the metric d_1 defined by

$$d_1(x, y) := |x_1 - y_1| + |x_2 - y_2|,$$

where $x = (x_1, x_2)$ and $y = (y_1, y_2)$.

From the point of view of convergence, there is little to choose between d_1 and d_E: It is easy to see that for all $x, y \in \mathbb{R}^2$,

$$d_1(x, y) \leq 2d_E(x, y),$$

$$d_E(x, y) \leq \sqrt{2}\, d_1(x, y)$$

[or even $d_1(x, y) \leq \sqrt{2}\, d_E(x, y)$, $d_E(x, y) \leq d_1(x, y)$]. Then d_1-convergence is the same as d_E-convergence. Geometrically, the metrics differ: d_1-balls do not look like d_E-balls. (See Exercise 5.E.)

(v) (An unorthodox metric on the integers.)

Let $x, y \in \mathbb{Z}$. If $x = y$, we put $d(x, y) = 0$. Otherwise, there is a positive integer m such that $x - y$ is divisible by 10^{m-1} but not by 10^m; then we set $d(x, y) := m^{-1}$.

Thus,

$$d(123, 4623) = \tfrac{1}{3},$$

$$d(10, 0) = \tfrac{1}{2},$$

$$d(3, 7) = 1,$$

$$d(3, -7) = \tfrac{1}{2}.$$

Observe that for all $x, y \in \mathbb{Z}$ and $p \in \mathbb{N}$,

$$d(x, y) < p^{-1} \iff 10^p \text{ divides } x - y.$$

This d really is a metric. Verification of Axioms M1, M2 and M3 is no problem; let us look into the Triangle Inequality. Let $x, y, z \in \mathbb{Z}$; we prove $d(x, z) \leq d(x, y) + d(y, z)$. We may assume $d(x, z) \neq 0$. If $d(x, z) = p^{-1}$, then 10^p does not divide $x - z$; hence, it cannot divide both $x - y$ and $y - z$. Then, $d(x, y) \geq p^{-1}$ or $d(y, z) \geq p^{-1}$. In either case, we obtain the desired inequality.

We call d the 10-*adic metric* on $\mathbb{Z}$.

To see just how unconventional this metric is, observe that (relative to d-convergence)

$$1.1! + 2.2! + 3.3! + \cdots = -1.$$

For a proof of this remarkable formula, put

$$x_n := 1.1! + 2.2! + \cdots + n.n! \qquad (n \in \mathbb{N}).$$

Inductively, one easily finds $x_n + 1 = (n+1)!$ For every $p \in \mathbb{N}$, we see that if n is large enough, $x_n + 1$ is divisible by 10^p, so that $d(x_n, -1) < p^{-1}$. Consequently, $x_n \to -1$.

5.6

The following definitions are straightforward generalizations of earlier ones.

Let (X, d) and (X', d') be metric spaces; let $f : X \to X'$. For $a \in X$, we say that f is *continuous at a* if

for every $\varepsilon > 0$,

there exists a $\delta > 0$ such that

$$d(x, a) < \delta \implies d'\big(f(x), f(a)\big) < \varepsilon.$$

f is simply called *continuous* if it is continuous at every point of X. (Occasionally, one may have to speak of "d-d'-continuity.")

We also have sequential continuity. We call f *sequentially continuous at a* if

$$x_n \to a \text{ in } X \implies f(x_n) \to f(a) \text{ in } X'.$$

Again, sequential continuity is the same as continuity:

Theorem 5.7
Let (X, d) and (X', d') be metric spaces. Let $f : X \to X'$ and $a \in X$. Then (α) and (β) are equivalent:

(α) f is continuous at a.
(β) f is sequentially continuous at a.

Proof
(precisely as in Theorems 1.12 and 3.20):
$(\alpha) \implies (\beta)$ Let $x_1, x_2, \ldots \in X$ and $x_n \to a$. Take $\varepsilon > 0$. Let δ be as in the definition of continuity in 5.6. For large n, we have $d(x_n, a) < \delta$; hence, $d'\big(f(x), f(a)\big) < \varepsilon$.
$(\beta) \implies (\alpha)$ Let $\varepsilon > 0$. We need a positive δ such that

$$x \in X, \ d(x, a) < \delta \implies d'\big(f(x), f(a)\big) < \varepsilon.$$

Suppose such a δ does not exist. Then, for every positive number δ there is an x in X with

$$d(x, a) < \delta \text{ but } d'\big(f(x), f(a)\big) \geq \varepsilon.$$

In particular, for every $n \in \mathbb{N}$ there is an $x_n \in X$ with

$$d(x_n, a) < n^{-1} \text{ but } d'\big(f(x_n), f(a)\big) \geq \varepsilon.$$

Then $x_n \to a$ but not $f(x_n) \to f(a)$. *Contradiction.* ∎

5.8
From this or from the definition of continuity, one easily proves: *If (X, d), (X', d'), and $(X, ''d'')$ are metric spaces and $f : X \to X'$ and $g : X' \to X''$ are continuous, then so is the composite map $g \circ f$.*

Examples 5.9
(X, d) and (X', d') are metric spaces.

(i) If X carries the trivial metric [5.3(vi)], then every map $X \to X'$ is continuous.

(ii) Let $f : X \to X'$. We say that f is *Lipschitz* (or *satisfies a Lipschitz condition*) if there exists a number K such that

$$d'\big(f(x), f(y)\big) \leq K d(x, y) \qquad (x, y \in X).$$

Such a K is called a *Lipschitz constant* for f.
Every Lipschitz map is continuous.

(iii) In particular, every isometry is continuous.

(iv) For every $a \in X$, the function

$$x \mapsto d(x, a) \qquad (x \in X)$$

is Lipschitz with Lipschitz constant 1 and therefore is continuous [Second Triangle Inequality, 5.4(i).]

(v) Let $X = X' = \mathbb{Z}$ and let d and d' be the 10-adic metric of 5.5(v). Then for every $a \in \mathbb{Z}$, the maps $x \mapsto x+a$ and $x \mapsto xa$ are Lipschitz (with Lipschitz constant 1), hence continuous.

(vi) Let d be the metric on $\mathcal{P}(\mathbb{N})$ described in 5.3(v). Let $D \in \mathcal{P}(\mathbb{N})$. For all $A, B \in \mathcal{P}(\mathbb{N})$,

$$d(A \cap D, B \cap D) \leq d(A, B).$$

(Check this.) Therefore, the map $A \mapsto A \cap D$ of $\mathcal{P}(\mathbb{N})$ into $\mathcal{P}(\mathbb{N})$ is Lipschitz, hence continuous. (So is the map $A \mapsto A \cup D$.)

Extra: Camille Jordan (1838–1922)

Gutenberg invented the printing press, Bell invented the telephone, and Herodotus is known as the father of history. Who invented topology?

If you have made it this far into the book, you know that there is no simple answer to the question. Topology, as a discipline on its own, emerged between two worlds, the one of Newton and the one of modern mathematics. At the divide, a complex pattern with various influences was woven and many players were at work. One of them was the Frenchman Camille Jordan. Born at Lyons, he was admitted to the École Polytechnique (School of Engineering) in Paris at age $17\frac{1}{2}$. On his entrance committee we find the famous Charles Hermite, incidentally one of the few great nineteenth-century French mathematicians who themselves had not been educated at the École Polytechnique. After his studies, Jordan chose to be an engineer for the French railroad, which somehow left him ample time for mathematical research. His interests were very wide and included probability theory, bilinear and quadratic forms, crystallography, analysis, and group theory. Here are some of his accomplishments that are not too far away from the subject of this book.

• He was the first to give a formal definition of a "curve" in the plane (1887). It is not without interest that his definition quickly and quietly was accepted, which should be contrasted with the very slow acceptance of Dirichlet's definition of a "function." In fact, none of his contemporaries bothered to credit Jordan for the definition, which may indicate that it was at hand but had simply not been written down before. (Various special curves had already been studied by other mathematicians.)

In Jordan's time, by the way, a plane curve was never represented by one letter, indicating a map from an interval to the plane, but by two, indicating the coordinate functions. In 1906, Maurice Fréchet in his doctoral thesis introduced the abstract notion of a metric space and thereby opened the road to our one-letter notation for a curve.

Jordan's name remains in the Jordan-Brouwer Separation Theorem and, indeed, in his Jordan curves.

• Jordan has priority to the attempted proof of what is now called the Jordan Closed Curve Theorem. Not only did this require him to think of totally new notions like domain and connectedness, but, more importantly than such technicalities, the effort shows how much mathematics had progressed in the nineteenth century. Previously, we compared Jordan's definition of a continuous curve with Dirichlet's definition of a function. Here, we may draw the analogy with Bolzano's desire to prove the Intermediate Value Theorem.

Even today, Jordan's Closed Curve Theorem is considered to be a deep theorem and all elementary proofs (like ours) are more or less technical at the least. A later branch of Topology, Algebraic Topology, has made Jordan's Closed Curve Theorem a simple consequence of much more general but equally technical results.

• Euler had pronounced the following result about polyhedra. Let P be a polyhedron. If one calls F the number of faces of P, V its number of vertices, and E its number of edges, one gets

$$2 = F + V - E.$$

(Try it out for a cube.) In *L'enseignement mathématique* ("The Teaching of Mathematics"), Lebesgue talks about *ce malheureux théorème d'Euler*, which we do not know to translate better than *that unpropitious theorem by Euler*. Unpropitious, because it took many generations of mathematicians to discover what exactly a polyhedron is. The final definition and theorem are due to Poincaré, but Lebesgue credits a lengthy memoir from 1866 by Jordan with the first correct proof of Euler's statement; and many authors today speak of the Euler-Jordan-Poincaré Theorem. We urge the reader to read the breathtaking book by Lakatos, *Proofs and Refutations*, which analyzes the historical road of that theorem every step of the way.

Jordan's contributions do not stop there. He was the first to attempt to define a measure of sets and he introduced and studied functions of bounded variation, which no doubt you will encounter in another course. Consequently, his name is honored with the Jordan Decomposition Theorem of Measure Theory. By no means were his results restricted to Analysis only. In Group Theory, we find the Jordan-Hölder Theorem and in Algebra you may have encountered Jordan's Canonical Form. Jordan made Galois' work accessible to a wide audience and, at the turn of the last century, his *Traité des substitutions* (1870) was as much a standard in Group Theory as his *Cours d'Analyse* was in analysis.

Jordan followed Hermite as professor in Analysis at the École Polytechnique in 1873. It comes as no surprise that he (with Riemann and Weierstrass) is considered one of the superteachers of the nineteenth century. Jordan remained active at the Collège de France where he had held a position concurrent with his professorship at the École Polytechnique. When Jordan passed away at age 84, the world of mathematicians had changed and Jordan had been one of the changers. He, perhaps, had not been of the genius quality of Riemann before him nor did he possess the all encompassing knowledge of the younger Poincaré, but his name became attached to as many as five beautiful theorems and he had influenced generations of mathematicians to come.

Further Reading

Lebesgue, H., Camille Jordan, *L'enseignement mathématique*, Sér. 2, 3, 1957, 81–106. (In French.)

Wilder, R.L., The Origin and Growth of Mathematical Concepts, *Bulletin of the American Mathematical Society*, 59, 1953, 423-448.

Young, L., *Mathematicians and their Times*, North-Holland, Amsterdam, 1981.

Exercises

5.A. Show that in the metric space of Example 5.5(v):
  (a) $\lim_{n \to \infty} 10^n = 0$.
  (b) $\lim_{n \to \infty} 2^n$ does not exist.

(c) $\displaystyle\sum_{n=1}^{\infty} 9 \cdot 10^n = -1$.

5.B. Consider $\mathbb{N} \times \mathbb{N}$ as a metric space under the Euclidean metric. Sketch the ball with center $(1, 1)$ and radius $\frac{5}{2}$ and the ball with center $(2, 2)$ and radius $\frac{3}{2}$. Show that the former is a proper subset of the latter.

5.C. Show that in $\mathbb{R}^2$, every ball contains a point P with rational coordinates and also a point Q with irrational coordinates.

5.D. Let d be the metric on $\mathcal{P}(\mathbb{N})$ as defined in Example 5.3(v). For $n = 1, 2, \ldots$, let

$$X_n := \{1^n, 2^n, 3^n, 4^n, \ldots\}.$$

Show that in the sense of d the sequence $X_1, X_2, \ldots$ converges.

5.E. (i) In 5.5(iv) we compared the metrics d_1 and d_E on $\mathbb{R}^2$. Show that the formula

$$d_\infty(x, y) := |x_1 - y_1| \vee |x_2 - y_2| \quad \left(x = (x_1, x_2),\ y = (y_1, y_2)\right)$$

(with " $\vee$ " as in 2.4) defines a metric d_∞ on $\mathbb{R}^2$ and that

$$d_\infty(x, y) \le d_E(x, y) \le \sqrt{2}\, d_\infty(x, y) \qquad (x, y \in \mathbb{R}^2).$$

Deduce that d_∞-convergence is d_E-convergence.

(ii) Sketch the sets

$$\{x : d_E(x, 0) < 1\},$$
$$\{x : d_1(x, 0) < 1\},$$
$$\{x : d_\infty(x, 0) < 1\}.$$

Another metric on $\mathbb{R}^2$ is the trivial metric, d_0. [See Example 5.3(vi).] Sketch

$$\{x : d_0(x, 0) < 1\}.$$

5.F. Consider $\mathbb{Z}$ with the 10-adic metric [Example 5.5(v)]. Let $x_n \to a$ and $y_n \to b$. Prove $x_n + y_n \to a + b$ and $x_n y_n \to ab$. [Example 5.9(v) might be of use.]

5.G. Let d be a metric in a set X. Prove that for all $a, b, x, y \in X$,

$$|d(x, y) - d(a, b)| \le d(x, a) + d(y, b).$$

Show

$$x_n \overset{d}{\to} a, \quad y_n \overset{d}{\to} b \quad \Longrightarrow \quad d(x_n, y_n) \to d(a, b).$$

5.H. Let (X', d') and $(X,'' d'')$ be metric spaces; let $X := X' \times X''$. For $x \in X$ we denote by x' and x'' its coordinates; thus: $x = (x', x'')$ with $x' \in X'$ and $x'' \in X''$.

For $x, y \in X$, put

$$d(x, y) := d'(x', y') + d''(x,'' y'').$$

(i) Prove that d is a metric on X; it is called the *sum-metric*.

(ii) Prove that, for $a, x_1, x_2, \ldots \in X$,

$$x_n \overset{d}{\to} a \quad \Longleftrightarrow \quad x_n' \overset{d'}{\to} a' \text{ and } x_n'' \overset{d''}{\to} a.''$$

5.I. Let (X, d) be a metric space. Define $d^* : X \times X \to [0, \infty)$ by

$$d^*(x, y) := d(x, y) \wedge 1.$$

(" $\wedge$ " as in 2.4.) Show that d^* is a metric on X and that for $a, x_1, x_2, \ldots \in X$,

$$x_n \overset{d}{\to} a \quad \Longleftrightarrow \quad x_n \overset{d^*}{\to} a.$$

Thus, *for every metric there exists a bounded metric that determines the same convergence.*

5.J. Let (X, d) be a metric space.
  (i) For $x, y \in X$, define

$$d'(x, y) := \sqrt{d(x, y)}.$$

  Is d' necessarily a metric?
  (ii) Same question for

$$d'(x, y) = \Big(d(x, y)\Big)^2.$$

5.K. Let (X, d) be a metric space and A a nonempty subset of X. Define $f_A : X \to \mathbb{R}$ by

$$f_A(x) := \inf\{d(x, a) : a \in A\}.$$

Intuitively, f_A is the distance from x to A:

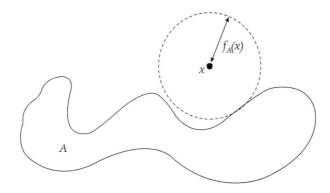

[5.9(iv) is the special case in which A consists of one element.] Show that

$$f_A(x) \leq f_A(y) + d(x, y) \qquad (x, y \in X).$$

Show that f_A is *Lipschitz with Lipschitz constant* 1.

5.L. *If h is a continuous function on $[0,1]$ and $\int_0^1 |h(x)|dx = 0$, then $h(x) = 0$ for all $x \in [0, 1]$. Indeed, define $H(t) = \int_0^t |h(x)|dx \ (0 \leq t \leq 1)$. As $H'(x) =$*

$|h(x)| \geq 0$, H is increasing; but $H(1) = 0 = H(0)$, so H must be constant. Then $|h(x)| = H'(x) = 0$ and $h(x) = 0$ for all x.

Now the exercise.

By $C[0, 1]$ we denote the set of all continuous functions on $[0, 1]$. For $f, g \in C[0, 1]$, define

$$d_1(f, g) := \int_0^1 |f(x) - g(x)| dx.$$

(i) Show that d_1 is a metric on $C[0, 1]$.

(ii) Is the map

$$f \mapsto \int_0^1 f(x) dx$$

a d_1-continuous function $C[0, 1] \to \mathbb{R}$?

(iii) Show that the function

$$f \mapsto f(1)$$

is *not* d_1-continuous. (Consider $f_n(x) = x^n$ ($x \in [0, 1]$, $n \in \mathbb{N}$) and prove that the sequence $(f_n)_{n \in \mathbb{N}}$ d_1-converges to the zero function.)

6

CHAPTER

Open and Closed Sets

In this chapter, (X, d) is a metric space.

Subsets of a Metric Space

6.1

We return to $\mathbb{R}$ and convergence of sequences in $\mathbb{R}$ for a moment. We have seen in the Connectedness Theorem (2.17) that for every nonempty subset A of $\mathbb{R}$ whose complement is also nonempty, there exists a number that is adherent to both A and $\mathbb{R} \backslash A$. The set of all such points is called the boundary of A, denoted

$$\partial A.$$

If a number is in ∂A, then it also is in $\partial(\mathbb{R} \backslash A)$, and vice versa. Thus,

$$\partial A = \partial(\mathbb{R} \backslash A).$$

Using Lemma 2.16, we have a convenient description of the boundary: A number c lies in the boundary of A if and only if there is a sequence in A tending to c and a sequence in $\mathbb{R} \backslash A$ tending to c as well.

Thus, 0 belongs to the boundary of the interval $(0, \frac{1}{2})$ because $n^{-1} \to 0$ and $-n^{-1} \to 0$. Similarly, $\frac{1}{2}$ is in the boundary of $(0, \frac{1}{2})$. The boundaries of $\mathbb{N}$, $\mathbb{Q}$, and $\mathbb{R} \backslash \{0\}$ are $\mathbb{N}$, $\mathbb{R}$, and $\{0\}$, respectively. [For $\mathbb{Q}$, observe that by Theorem 2.8(i) every number is the limit of some sequence $(x_n)_{n \in \mathbb{N}}$ of rationals; then it is also the limit of the irrational numbers $x_n + n^{-1}\sqrt{2}$.]

6.2

We can define analogous notions in any metric space (X, d). Let $A \subset X$, $c \in X$. We say that c is *adherent to A* if $B_\varepsilon(c) \cap A \neq \emptyset$ for every $\varepsilon > 0$. [Recall that $B_\varepsilon(a)$ is the ball on a with radius ε; see 5.4(ii).]

c is said to be *in the boundary* of A if it is adherent to both A and $X \backslash A$. The set of all points that are in the boundary of A is called the *boundary* of A, denoted

$$\partial A.$$

Here too, we have

$$\partial A = \partial(X \backslash A).$$

6.3

Let $A \subset X$. By definition, A is *open in X* if $\partial A \subset X \backslash A$ and A is *closed in X* if $\partial A \subset A$. These definitions may seem somewhat mysterious but they will be clarified in Theorem 6.6.

A point c of X is said to be *interior to A* if there exists an $\varepsilon > 0$ such that $B_\varepsilon(c) \subset A$.

6.4

The observations (i)-(iii) are immediate.

(i) If $A_1 \subset A_2 \subset X$ and $c \in X$, then

$$c \text{ is interior to } A_1 \quad \Longrightarrow \quad c \text{ is interior to } A_2$$

and

$$c \text{ is adherent to } A_1 \quad \Longrightarrow \quad c \text{ is adherent to } A_2.$$

(ii) If $A \subset X$ and $c \in X$, then

> c *is interior to* $A \Longrightarrow c \in A,$
>
> $c \in A \Longrightarrow c$ *is adherent to* $A.$

(iii) If $A \subset X$ and $c \in X$, then

> c *is interior to* $A \iff c$ *is not adherent to* $X \backslash A,$
>
> c *is adherent to* $A \iff c$ *is not interior to* $X \backslash A.$

From observation (iii) one sees that ∂A consists precisely of those points that are interior to neither $X \backslash A$ nor A. Therefore:

Theorem 6.5
If $A \subset X$, then X is the union of the following three sets that are pairwise disjoint (but may be empty):

$$\left[\begin{array}{l} \text{the set of points interior to } A, \\ \text{the set of points interior to } X \backslash A, \\ \text{the boundary of } A. \end{array}\right.$$

It follows that our definitions are more symmetric than they appear at first glance. This is also reflected in

Theorem 6.6
If A and B are subsets of X that are each other's complements, then

$$A \text{ is closed in } X \iff B \text{ is open in } X.$$

We leave the proof as an exercise.

The following facts tie the above notions together. Again, we leave the proof to the reader.

Theorem 6.7

(i) *For $A \subset X$, (α)-(δ) are equivalent.*
    (α) *A is open.*
    (β) *$\partial A \subset X \backslash A$.*
    (γ) *Every point of A is interior to A.*
    (δ) *For every $c \in A$ there exists an $\varepsilon > 0$ with $B_\varepsilon(c) \subset A$.*
(ii) *For $A \subset X$, (α)-(δ) are equivalent.*
    (α) *A is closed.*
    (β) *$\partial A \subset A$.*
    (γ) *Every point of X that is adherent to A lies in A.*
    (δ) *If $c \in X$ and $B_\varepsilon(c) \cap A \neq \varnothing$ for all $\varepsilon > 0$, then $c \in A$.*

6.8
Let A be a subset of X. It is straightforward to prove a generalization of Lemma 2.16: *A point c of X is adherent to A if and only if A contains a sequence that converges to c.* This observation leads to an alternative description of closedness:

A subset A of a metric space X is called *sequentially closed* if for every sequence in A that converges in A the limit is in A, i.e., if

$$\left.\begin{array}{c} x_1, x_2, \ldots \in A \\ c \in X \\ x_n \to c \end{array}\right] \implies c \in A.$$

We see immediately that *every closed set is sequentially closed and vice versa*. (Like sequential continuity, sequential closedness seems an entirely redundant concept. Further on we will show our reasons for introducing it.)

Examples 6.9

(i) Consider the metric space $\mathbb{R}$. (Euclidean metric.) If $c \in (0, 1)$ and if for ε we take the smallest of the numbers c and $1-c$, then $\varepsilon > 0$ and $B_\varepsilon(c) \subset (0, 1)$. Thus, $(0, 1)$ is open. $[0, 1]$ is sequentially closed, hence closed.

(ii) At the end of 6.1 we have seen that in the metric space $\mathbb{R}$, the boundary of $\mathbb{Q}$ is $\mathbb{R}$ itself. Thus, we have neither $\partial\mathbb{Q} \subset \mathbb{R}\backslash\mathbb{Q}$ nor $\partial\mathbb{Q} \subset \mathbb{Q}$. This means that $\mathbb{Q}$ is not open, but not closed either. We see that *subsets of a metric space may be neither open nor closed*. On the other hand, *they may be both open and closed*, as we will see in the next example.

(iii) Consider the trivial metric on a set X. [See 5.3(vi).] Then $B_1(a) = \{a\}$ for all $a \in X$. Consequently, every point of a subset A of X is interior to A, but also, every point that is adherent to A lies in A. Thus, every subset of X is both open and closed.

(iv) In the metric space $\mathbb{R}$, no point is interior to $\mathbb{Z}$, so $\mathbb{Z}$ is not open in $\mathbb{R}$. A point of $\mathbb{R}$ that is adherent to $\mathbb{Z}$ is an element of $\mathbb{Z}$, so $\mathbb{Z}$ is closed in $\mathbb{R}$.

(v) The boundary of the whole set X is $\varnothing$. In our initial discussion in 6.1, we did not allow A to be all of $\mathbb{R}$, but you might have noticed that there was no restriction on A in our *definition* of "boundary." As a consequence, we get that X is both open and closed in X.

(vi) Since there is no restriction on A, we also allow A to be $\varnothing$. This may puzzle you for a moment, but if you apply the definitions ruthlessly, you will find that $\partial\varnothing = \varnothing$ and that $\varnothing$ is both open and closed. (If this juggling with the empty set worries you, read 6.27.)

(vii) If (X, d) is a metric space and $a \in X$, then the set $\{a\}$ is closed.

(viii) Every interval of the form $[a, b]$ is (sequentially) closed in $\mathbb{R}$. Conversely, the only bounded intervals that are closed in $\mathbb{R}$ *are* of the type $[a, b]$. Our terminology of "closed" coincides with the earlier use of that word in "closed bounded interval."

Also, the "closed unit disk" and the "closed unit square" (see Example 4.19) are closed subsets of $\mathbb{R}^2$.

Example 6.10
For most of the subsets of metric spaces that we have encountered so far, you will have no problem determining their boundaries. We will now

construct a subset of $\mathbb{R}$ that is a little more complicated. The construction proceeds in infinitely many steps.

In step 1, we take the interval $[0, 1]$ and leave off the subinterval $(\frac{1}{3}, \frac{2}{3})$. The remainder is the union of two intervals, $[0, \frac{1}{3}]$ and $[\frac{2}{3}, 1]$. In step 2, we leave off the middle third of each of these. We are left with four intervals of length $\frac{1}{9}$:

$$[0, \tfrac{1}{9}], [\tfrac{2}{9}, \tfrac{1}{3}], [\tfrac{2}{3}, \tfrac{7}{9}], [\tfrac{8}{9}, 1].$$

In this way, in the n-th step, we obtain 2^n intervals of length 3^{-n}; their union we call C_n. Trivially,

$$[0, 1] \supset C_1 \supset C_2 \supset \cdots$$

The set we want to look at is the intersection

$$C := \{x \in \mathbb{R} : x \in C_n \text{ for every } n\},$$

called the *Cantor set*.

Observe that C is not empty: 0 and 1 lie in every C_n, hence also in C. So do $\frac{1}{3}$ and $\frac{2}{3}$, and the end points of the four intervals that constitute C_2, and so forth. We see that C actually is an infinite set.

It is obvious that C_2 does not contain any interval of length more than $\frac{1}{9}$; then neither does C. Similarly, C does not contain any interval of length more than 3^{-n} for any $n \in \mathbb{N}$. But then C does not contain any interval at all. Thus, *no point of $\mathbb{R}$ is interior to C.*

One easily sees that each C_n is closed. It follows that C is *closed.* Indeed, let $c \in \mathbb{R}$ be adherent to C. For every n, we have $C \subset C_n$, so that c is adherent to C_n, from which $c \in C_n$. Then $c \in C$ by the definition of C.

We note that the preceding argument applies to every sequence of closed sets: If $A_1, A_2, \ldots$ are closed, then so is $A := \{x : x \in A_n \text{ for every } n\}$. Indeed, if c is adherent to A, then it is adherent to every A_n because $A \subset A_n$; then $c \in A_n$ for every n; then $c \in A$. You can see that this has nothing to do with the closed sets forming a sequence. The same idea works for *any* collection of closed sets.

In behalf of the reader who is not familiar with sets of sets we digress for a moment to introduce some notation and terminology.

Collections of Sets

6.11

In itself there is nothing particularly troublesome about sets of sets, but you have to watch them.

You will have no problem with specific instances such as the set of all intervals, say, or the set of all circles in the plane. The elements of such a set are themselves sets (intervals, circles) and consist of elements (numbers, points of the plane). But the terminology gets confusing: Whenever you use the word "element", you have to realize very clearly on which level you are working.

To prevent confusion, we adapt our terminology. We avoid talking of "a set of sets" but say "a collection of sets," such as "the collection of all circles." We do not say that a given circle "is an element" of this collection, but that it "belongs to" the collection. We use Greek letters ($\alpha, \omega, \varphi, \ldots$) to indicate collections. If we want to express the fact that a set A belongs to a collection α, we do not write "$A \in \alpha$" but

$$A \mathrel{\unicode{x22F1}} \alpha$$

and instead of "$\alpha \subset \beta$" (where α and β are collections of sets), we write

$$\alpha \mathrel{\sqsubset} \beta.$$

It should be understood that these notations do not make any substantial difference but are merely stylistic devices. The formula "$A \mathrel{\unicode{x22F1}} \alpha$" does not really mean anything else than "$A \in \alpha$," but it is intended to recall to the reader that A is a set.

It is not always useful to do so. Sometimes, the nature of the object A is irrelevant and it may be distracting to keep stressing that A is a set. Actually, we have already been using formulas like "$A \in \alpha$." In Example 5.3(v) we considered $\mathcal{P}(\mathbb{N})$, the set of all subsets of $\mathbb{N}$, and observed that the set P of all prime numbers is an element of $\mathcal{P}(\mathbb{N})$. In our new terminology, we could call $\mathcal{P}(\mathbb{N})$ the "collection" of all subsets of $\mathbb{N}$ and write "$P \mathrel{\unicode{x22F1}} \mathcal{P}(\mathbb{N})$" to indicate that P "belongs to" $\mathcal{P}(\mathbb{N})$. However, when dealing with $\mathcal{P}(\mathbb{N})$ as a metric space, we will prefer to consider P as a point in $\mathcal{P}(\mathbb{N})$ and ignore its internal structure.

6.12

Let Y be a set and β a collection of subsets of Y. Assume β is nonempty, i.e., there is a subset of Y that belongs to β. The *union* of β,

$$\bigcup \beta,$$

is the set

$$\{y \in Y : y \in B \text{ for some } B \mathrel{\unicode{x22F1}} \beta\},$$

which is a subset of Y. Similarly, the *intersection* of β,

$$\bigcap \beta,$$

is

$$\{y \in Y : y \in B \text{ for every } B \in \beta\}.$$

For instance, let Y be $\mathbb{R}$ and let β be the collection of all intervals $[t, t+3]$ where $t \in [0, 1]$. Then

$$y \in \bigcup \beta \iff \text{there is a } t \in [0, 1] \text{ with } y \in [t, t+3]$$

and

$$y \in \bigcap \beta \iff \text{for all } t \in [0, 1] \text{ we have } y \in [t, t+3].$$

Thus, $\bigcup \beta$ is $[0, 4]$ and $\bigcap \beta$ is $[1, 3]$.

6.13
In the above we have required β to be nonempty. To avoid complications further on, it is worthwhile to have a look at $\bigcup \beta$ and $\bigcap \beta$ for $\beta = \varnothing$.

As you can see, it is perfectly in line with the above to define

$$\bigcup \varnothing := \varnothing.$$

"$\bigcap \varnothing$" is less straightforward. For nonempty β, we have

$$\bigcap \beta = \{y \in Y : y \in B \text{ for every } B \in \beta\}.$$

Given an element y of Y, we might say that every $B \in \beta$ subjects y to a test ("Is y an element of B?"); if y passes all these tests, it gets membership of the set $\bigcap \beta$. The smaller β is, the fewer tests have to be passed and the larger $\bigcap \beta$ will be. In the case $\beta = \varnothing$, there are no tests at all and every y is automatically admitted to $\bigcap \beta$:

$$\bigcap \varnothing = Y.$$

It seems paradoxical that $\bigcap \varnothing$ should be larger than $\bigcup \varnothing$, but that is what happens if for $\beta = \varnothing$, we stick to the same definition we used for $\beta \neq \varnothing$. It is the definition of "$\bigcap \varnothing$" that is accepted in set theory. The real trouble with it is that it depends on Y. We cannot say what $\bigcap \varnothing$ is as long as we do not specify our "universe" Y.

In the case of a nonempty β, such a problem does not occur. In 6.12 we have observed that $\bigcup \beta = [0, 4]$ and $\bigcap \beta = [1, 3]$ if $Y = \mathbb{R}$ and $\beta = \{[t, t+3] : 0 \le t \le 1\}$, but, trivially, $\bigcup \beta$ and $\bigcup \beta$ do not change if we replace $\mathbb{R}$ by any other set Y (as long as β is a collection of subsets of Y, of course).

We choose the coward's way out. We will not use $\bigcap \varnothing$.

6.14
As before, (X, d) is a metric space.

Picking up the thread we left at the end of Example 6.10, we observe that the intersection of any nonempty collection α of closed subsets of X is closed. Indeed, let $c \in X$ be adherent to $\bigcap \alpha$. Then, for every $A \in \alpha$, c is adherent to A (simply because $\bigcap \alpha \subset A$), so $c \in A$; thus, $c \in \bigcap \alpha$ by the definition of $\bigcap \alpha$.

With *unions* of closed sets we have to be more careful. We know that in $\mathbb{R}$, every subset with only one element is closed, and every subset of $\mathbb{R}$ is a union of one-element subsets. But not every subset of $\mathbb{R}$ is closed. Hence, not every union of closed sets is closed.

On the other hand, the union of a nonempty collection ω of *open* sets is *open*: Let $c \in \bigcup \omega$. Then there is a $W \in \omega$ with $c \in W$; thus, c is interior to W, hence to $\bigcup \omega$ (because $W \subset \bigcup \omega$).

How about intersections of open sets? In $\mathbb{R}$, the intersection of the open sets $(-n^{-1}, n^{-1})$ ($n \in \mathbb{N}$) is the set $\{0\}$, which certainly is not open. However, any intersection of *two* open sets is open. To see this, take open sets W_1 and W_2, and let $c \in W_1 \cap W_2$: We wish to prove that c is interior to $W_1 \cap W_2$. As $c \in W_1$ and $c \in W_2$, there exist $\varepsilon_1 > 0$ with $B_{\varepsilon_1}(c) \subset W_1$ and $\varepsilon_2 > 0$ with $B_{\varepsilon_2}(c) \subset W_2$. If ε is the smaller of the numbers ε_1 and ε_2, then $B_\varepsilon(c) \subset B_{\varepsilon_1}(c) \subset W_1$ and $B_\varepsilon(c) \subset B_{\varepsilon_2}(c) \subset W_2$, so $B_\varepsilon(c) \subset W_1 \cap W_2$.

With this result, we can see that a union of *two* closed sets is always closed: If A_1 and A_2 are closed, then $X \backslash A_1$ and $X \backslash A_2$ are open, hence so is their intersection. But

$$(X \backslash A_1) \cap (X \backslash A_2) = X \backslash (A_1 \cup A_2). \qquad (*)$$

(Check this.) We see that the complement of $A_1 \cup A_2$ is open. Then $A_1 \cup A_2$ itself is closed.

We collect these results in a theorem:

Theorem 6.15

Let (X,d) be a metric space.

(I)   (i) *The intersection of any two open subsets of X is open in X.*
        (ii) *The union of any nonempty collection of open subsets of X is open in X.*

(II)   (i) *The union of any two closed subsets of X is closed in X.*
        (ii) *The intersection of any nonempty collection of closed subsets of X is closed in X.*

6.16

Look back at our proof of the above theorem. We have proved (II)(i) as a consequence of (I)(i), using the symmetry between openness and closedness and formula $(*)$ of 6.14. We can apply the same principle to derive

(II)(ii) from (I)(ii) (or contrariwise). Let us see how that is done, not because having two proofs is better than having only one, but because the technique of the second proof is of importance.

Assume (I)(ii). Let α be a nonempty collection of closed subsets of X. The sets $X \backslash A$ with $A \in \alpha$ form a nonempty collection ω of open sets. Then $\bigcup \omega$ is open, so $X \backslash \bigcup \omega$ is closed. But $X \backslash \bigcup \omega$ is precisely $\bigcap \alpha$. Indeed, for $x \in X$ we have

$$x \in X \backslash \bigcup \omega \iff x \notin \bigcup \omega$$

$$\iff \text{there is no } W \in \omega \text{ with } x \in W$$

$$\iff \text{for all } W \in \omega \text{ we have } x \notin W$$

$$\iff \text{for all } A \in \alpha \text{ we have } x \notin X \backslash A$$

$$\iff \text{for all } A \in \alpha \text{ we have } x \in A$$

$$\iff x \in \bigcap \alpha.$$

6.17

Notice that we have defined what it means for a set to be open or closed *in X*. Indeed, $[0, 1]$ is not open in $\mathbb{R}$, but it *is* open in the space $[0, 1] \cup [2, 3]$ (under the metric induced by the Euclidean one, of course). Thus, "open" is a relative notion and, at times, when we want to be precise, we will use "open in ..." rather than just "open."

Having said this, we admit that from now on we will drop the specification when the context makes it superfluous.

The same remarks can be made about the words "closed," "interior," "adherence," and "boundary," all of which refer to relative concepts. (For instance, as a subset of $\mathbb{R}$, the Cantor set has no interior points; as a subset of itself, it has many. In $\mathbb{R}$, the boundary of $[0, 1]$ is $\{0, 1\}$; in $[0, 1] \cup [2, 3]$, the boundary of $[0, 1]$ is empty.)

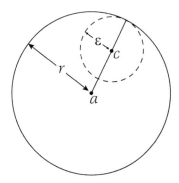

Lemma 6.18
Every ball in a metric space is open.

Proof

Let $a \in X, r > 0, c \in B_r(a)$; we wish to prove that c is interior to $B_r(a)$. That is easy. The number

$$\varepsilon := r - d(a, c)$$

is positive, and by the Triangle Inequality, $B_\varepsilon(c) \subset B_r(a)$. ∎

The proof of the following characterization of open sets is left as an exercise.

Theorem 6.19

A subset of a metric space is open if and only if it is the union of a collection of balls.

Similar Metrics

6.20

We return to convergence. For $a \in X$ and a sequence $(x_n)_{n \in \mathbb{N}}$ in X, the formula $\lim_{n \to \infty} x_n = a$ means

$$\text{Every ball with center } a \tag{1}$$
$$\text{contains a tail of the sequence.}$$

A ball is a particular example of an open set, so (1) is certainly true if

$$\text{Every open set containing } a \tag{2}$$
$$\text{contains a tail of the sequence.}$$

On the other hand, if U is an open set and $a \in U$, then $B_\varepsilon(a) \subset U$ for some ε. Hence, (1) also implies (2). Thus, (2) is equivalent to $\lim_{n \to \infty} x_n = a$.

These observations have the following consequences. Suppose d_1 and d_2 are metrics on the same set X. Let us call d_1 and d_2 *similar* if

$$x_n \xrightarrow{d_1} x \quad \Longleftrightarrow \quad x_n \xrightarrow{d_2} x.$$

[An example is given by the Euclidean and the $\tan^{-1}$ metric on $\mathbb{R}$; see Example 5.5(ii).]

From the above, we see that this is the case if the notions "d_1-open" and "d_2-open" coincide. The converse is also true. In fact, we have:

Theorem 6.21

For two metrics d_1 and d_2 on X, conditions (α), (β), and (γ) are equivalent.

(α) d_1*-convergence is d_2-convergence. (d_1 and d_2 are similar.)*
(β) *The d_1-open sets are precisely the d_2-open sets.*

(γ) *The d_1-closed sets are precisely the d_2-closed sets.*

Proof

We have just seen (β) $\implies$ (α). The closed sets are just the complements of the open sets; hence, (γ) $\implies$ (β). Third, (α) implies that d_1-sequential closedness is the same as d_2-sequential closedness. Therefore, (α) $\implies$ (γ). ∎

Suppose we have two sets, X and X', two similar metrics, d_1 and d_2, on X, and two similar metrics, d_1' and d_2', on X'. Let $f : X \to X'$. Continuity of a map between metric spaces is the same as sequential continuity. It follows that

$$f \text{ is } d_1\text{-}d_1'\text{-continuous} \iff f \text{ is } d_2\text{-}d_2'\text{-continuous.}$$

Then it ought to be possible to describe continuity of f without explicitly mentioning metrics, in terms of open sets or closed sets only. We proceed to do so.

6.22

First, we remind you of a notation. If $f : X \to X'$ is a map and $A \subset X'$, then

$$f^{-1}(A) := \{x \in X : f(x) \in A\}.$$

$f^{-1}(A)$ is called the *inverse image of A under f*. (Keep in mind that f is not required to be injective: There may not be a map f^{-1}.)

Let (X, d) and (X', d') be metric spaces and let $a \in X$ and $f : X \to X'$. We call f *continuous* at a if (see 5.6)

> for every $\varepsilon > 0$ there exists a $\delta > 0$
>
> such that $x \in B_\delta(a) \implies f(x) \in B_\varepsilon(f(a))$,

i.e., if

> for every ball B in X' with center $f(a)$,
>
> a is interior to $f^{-1}(B)$.

Again, balls are special examples of open sets, and continuity of f at a is equivalent to

> for every open $V \subset X'$ with $f(a) \in V$,
>
> a is interior to $f^{-1}(V)$.

and to

> for every open $V \subset X'$ with $f(a) \in V$,
>
> there is an open $U \subset X$ with $a \in U \subset f^{-1}(V)$.

(Check this.)

We now have our description of continuity in the language of open sets. The main results of 6.20 and 6.22 are the following:

Theorem 6.23
Let X and X' be metric spaces, $a \in X$.

(i) *A sequence in X converges to a if and only if every open set U in X with $a \in U$ contains a tail of the sequence.*
(ii) *$f:X \to X'$ is continuous at a if and only if for every open set V in X' with $f(a) \in V$, there exists an open U in X such that $a \in U \subset f^{-1}(V)$.*

The description of "global" continuity in terms of open sets is more elegant:

Theorem 6.24
Let X and X' be metric spaces. For $f:X \to X'$, conditions (α), (β), and (γ) are equivalent.

(α) *f is continuous.*
(β) *For every open $V \subset X'$, the set $f^{-1}(V)$ is open in X.*
(γ) *For every closed $A \subset X'$, the set $f^{-1}(A)$ is closed in X.*

The proof is left to the reader.

Interior and Closure

6.25
Before we end this chapter, we introduce some more terminology for future reference.

Let $Y \subset X$. The *closure of Y in X* is the set of all points of X that are adherent to Y; it is denoted clo(Y) or Y^- [or, if we are forced to it, X-clo(Y)]. The *interior of Y in X* is the set of all points of X that are interior to Y; it is denoted int(Y) or Y°. (Thus, $Y^{\circ-}$ will be the closure of the interior of Y.)

Theorem 6.26
Let $Y \subset X$.

(i) *Y^- is the smallest closed set containing Y, i.e., Y^- is closed; $Y^- \supset Y$; if A is closed and $A \supset Y$, then $A \supset Y^-$.*
(ii) *Y° is the largest open set contained in Y, i.e., Y° is open; $Y^\circ \subset Y$; if U is open and $U \subset Y$, then $U \subset Y^\circ$.*

We leave the proof to you.

The Empty Set

6.27
For readers who feel uncomfortable with $\varnothing$, the following discussion might be useful.

Consider the statement

If x is a real number and $x > 1$, then $3x > 2$.

Its truth seems to be unassailable. Now try $x = 2$. Certainly, 2 is a real number, so we get

$$\text{If } 2 > 1, \text{ then } 6 > 2. \tag{1}$$

Similarly, by choosing $x = 1$ and $x = 0$, we find

$$\text{If } 1 > 1, \text{ then } 4 > 2. \tag{2}$$

$$\text{If } 0 > 1, \text{ then } 0 > 2. \tag{3}$$

Are these conclusions valid? If you ask a nonmathematician, the answer probably will be that all three statements are ridiculous and the question of whether they are true is not applicable.

In mathematics the locution "if ..., then ... " has been given a more precise meaning than in daily life. If A and B are statements, then

$$\text{If } A, \text{ then } B \tag{$*$}$$

is held to be true in each of the following cases:

A is true, B is true;

A is false, B is true;

A is false, B is false.

Thus, mathematically, statements (1), (2), and (3) are accepted as being true. Note that the validity of $(*)$ does not require any intrinsic connection between A and B. For instance,

$$\text{If } 2 + 2 = 4, \text{ then } \pi > 0$$

is a true statement.

$(*)$ is true whenever B is true. Somewhat disturbingly, $(*)$ is true whenever A is false. This brings us to the empty set. We see now that, for $x \in \mathbb{R}$, we have

$$\text{If } x \in \varnothing, \text{ then } x > 0;$$

$$\text{If } x \in \varnothing, \text{ then } \sin x = \tfrac{1}{2};$$

etc.

In short, any sentence of the form "if $x \in \varnothing$, then $\mathcal{A}$" is necessarily true. (Provided that $\mathcal{A}$ represents a meaningful statement, "If $x \in \varnothing$, then $x \mathrel{>=}$ " is still ridiculous.)

As an example, let (X, d) be a metric space. For $x \in X$, we have

$$\text{If } x \in \varnothing, \text{ then } B_{\frac{1}{2}}(x) \subset \varnothing.$$

Consequently, $\varnothing$ is an open set. As $\varnothing$ clearly has no adherent points, every point of X that is adherent to $\varnothing$ lies in $\varnothing$. Hence, $\varnothing$ is also closed.

Extra: Cantor (1845–1918)

Georg Cantor worked all of his professional life at the university in Halle. Halle was not one of the prestigious European universities, but it was no lack of ambition on Cantor's side that he spent most of his life there. He had always wanted to work in a better recognized research center. Part of the blame can be put on his revolutionary insights. It is not easy to propel ideas that are fundamentally contrary to the accepted norm. Moreover, Cantor was brought up in a very religious environment and his personality could be described as a little rigid in dealing with other people's convictions. The resistance to his ideas was led by Leopold Kronecker (1823-1891). Still, in his lifetime, though, praise for his work started to mount. David Hilbert (1862-1943) expressed his admiration: "Cantor has created a paradise from which no one shall expel us." Cantor is the founding father of set theory and the first who defined irrational numbers by sequences of rational numbers. With Richard Dedekind (1831-1916) and Karl Weierstrass (1815-1897), he worked on the foundations of analysis as we now know it. The influence of his thoughts on twentieth-century mathematics is profound and lasting.

Cantor's first set theory paper was his 1874 paper on algebraic numbers. An algebraic number is any real number that is a zero of a nonconstant polynomial with integer coefficients. Cantor proved that the set of all algebraic numbers can be put in a one-to-one correspondence with the set of all positive integers. More importantly, he showed that the set of all real numbers cannot be put in such a one-to-one correspondence. In the words of Chapter 19, the set of real numbers is not equipollent with the set of all positive integers. Encouraged by these successes, he introduced the notion of equipollency of sets in 1878. Cantor proved that any $\mathbb{R}^N$ is equipollent to the set of all real numbers. He finishes the 1878 paper with the conjecture that became known as the Continuum Hypothesis (see Chapter 19).

The possibility of a one-to-one correspondence between an infinite set and one of its proper subsets was not a totally new idea. It had been observed earlier by Galileo, Leibniz, and Bolzano. The novelty of

Cantor's contributions is his courage. Where the previous authors had concluded a wasteland, he moved on to define infinite sets of equal power. A comparison with Galileo is not as wild as it might seem on first sight.

Only with Weierstrass, the mathematical world as a whole came to a generally accepted conclusion about a remnant of the olden days of Newton: infinitesimals. Accrediting the existence of an actual infinity has turned out to be equally important to the development of mathematics as the solidification of Calculus, begun by Cauchy and Bolzano and completed by Weierstrass.

To solve the Continuum Hypothesis was not within Cantor's reach. Growing increasingly frustrated about this and about the lack of general recognition, Cantor had a nervous breakdown in 1884. From then on, mental problems would remain part of his life. He was denied a professorship in Berlin, vehemently opposed by Kronecker. Meanwhile, outside Germany, Cantor's ideas were gaining support. The recognition of sets as a notion underlying all of mathematics led to many entirely new fields, of which topology is a good example. In spite of his mental illness, Cantor continued to work hard. He instituted the German Mathematical Society, founded in 1889, and was instrumental in establishing the first International Congress of Mathematicians (1897) in Zürich.

At the Second International Congress of Mathematicians in Paris (1900), the Continuum Hypothesis was first among 23 problems that Hilbert proposed as central to the development of twentieth-century mathematics. The controversy between Kronecker and Cantor resurfaced in a new battle about the foundations of mathematics. The debate was settled, in a way, by work of Kurt Gödel (in 1938) and Paul J. Cohen (in 1963).

Cantor died on January 6, 1918 at the psychiatric hospital in Halle.

Further Reading

Dauben, J.W., *Georg Cantor, His Mathematics and Philosophy of the Infinite*, Princeton University Press, Princeton, NJ, 1979.

Young, L., *Mathematicians and Their Times*, North-Holland Publishing Company, Amsterdam, 1981.

Exercises

6.A. The *graph* of a map $f : X \to Y$ is the set

$$\{(x, y) \in X \times Y : y = f(x)\}.$$

(i) Let X, Y be metric spaces. We endow the space $X \times Y$ with the sum-metric. (See Exercise 5.H.) Prove: If $f : X \to Y$ is continuous, then its graph is closed in $X \times Y$.

(ii) Define $f : [0, \infty) \to [0, \infty)$ by

$$f(x) := x^{-1} \quad \text{if} \quad x > 0,$$

$$f(0) := 0.$$

Show that the graph of f is closed in $[0, \infty) \times [0, \infty)$ (although, of course, f is not continuous).

6.B. We consider the set

$$X = \left\{ 1, \frac{1}{2}, \frac{1}{3}, \frac{1}{4}, \dots \right\} \cup \{0\}$$

as a metric space under the Euclidean metric.

(i) Prove: For every $n \in \mathbb{N}$, the singleton set $\{\frac{1}{n}\}$ is open in X; the set $\{0\}$ is not open in X.

(ii) Let $Y \subset X$. Prove: If $0 \notin Y$, then Y is open in X; if $0 \in Y$, then Y is closed in X. [Thus, every subset of X is open or closed (or both). Such a metric space is sometimes called a "door space."]

(iii) Let $f : X \to \mathbb{R}$. Prove:

$$f \text{ is continuous} \quad \Longleftrightarrow \quad \lim_{n \to \infty} f\left(\frac{1}{n}\right) = f(0).$$

6.C. Let Y be a subset of a metric space X. Prove that for $a \in X$,

$$a \in \partial Y \quad \Longleftrightarrow \quad \mathbf{1}_Y \text{ is not continuous at } a.$$

6.D. Let A be a subset of a metric space X. Prove

$$A \text{ is open and closed} \quad \Longleftrightarrow \quad \partial A = \varnothing.$$

6.E. (Sequel to 5.K.) Show that if A is closed,

$$f_A(x) = 0 \quad \Longleftrightarrow \quad x \in A.$$

6.F. Let f and g be continuous real-valued functions on a metric space X. Show that $\{x \in X : f(x) > g(x)\}$ is open.

6.G. (i) Let X be a set and α a nonempty collection of subsets of X. Let $T \subset X$ and let β be the collection of all sets of the form $A \cup T$ where A belongs to α. Show that

$$\bigcup \beta = \left(\bigcup \alpha\right) \cup T, \qquad \bigcap \beta = \left(\bigcap \alpha\right) \cup T.$$

Hint: Prove that for $x \in X$, one has $x \in \bigcup \beta \Longleftrightarrow x \in \left(\bigcup \alpha\right) \cup T$.

(ii) Part (i) was just a warm-up exercise for the following. Let f be a map of a set X into a set Y. Let β be a nonempty collection of subsets of Y and let α be the collection of all subsets of X of the form $f^{-1}(B)$, where B belongs to β. Show that

$$\bigcup \alpha = f^{-1}\left(\bigcup \beta\right), \qquad \bigcap \alpha = f^{-1}\left(\bigcap \beta\right).$$

(iii) Let f be a map of a set X into a set Y. Let σ be a collection of subsets of X and τ the collection of all sets $f(S)$ where S belongs to σ. Prove that

$$\bigcup \tau = f(\bigcup \sigma).$$

Give an example showing that *not always* $\bigcap \tau = f(\bigcap \sigma)$. (It is possible to make one with only two sets belonging to σ.)

6.H. Determine the interior, the closure, and the boundary of
  (a) the interval $[0, 1)$ in the metric space $\mathbb{R}$;
  (b) the interval $[0, 1]$ in the metric space $[0, 1]$;
  (c) the set $\{m+n\pi : m, n \in \mathbb{N}\}$ in $\mathbb{R}$;
  (d) the set $\{\frac{1}{m} + \frac{1}{n} : m, n \in \mathbb{N}\}$ in $\mathbb{R}$;
  (e) the set $\{(x, y) : 0 < x^2+y^2 < 1\}$ in $\mathbb{R}^2$;
  (f) the set $\{(x, 0) : x \in \mathbb{R}\}$ in $\mathbb{R}^2$;
  (g) the set $\{(m, n) : m, n \in \mathbb{N}\}$ in $\mathbb{R}^2$;
  (h) the set $\{(\frac{1}{m}, \frac{1}{n}) : m, n \in \mathbb{N}\}$ in $\mathbb{R}^2$.

6.I. Let (X, d) be a metric space, $a \in X$, and $r > 0$. Show that

$$\{x : d(x, a) < r\}^- \subset \{x : d(x, a) \le r\}.$$

Give an example in which the sets differ.

6.J. Check whether each of the following formulas holds for *all* metric spaces X, X', *all* continuous maps $f : X \to X'$, and *all* subsets A of X.
  (a) $f(A^-) \subset (f(A))^-$.
  (b) $f(A^-) \supset (f(A))^-$.
  (c) $f^{-1}(A^-) \subset (f^{-1}(A))^-$.
  (d) $f^{-1}(A^-) \supset (f^{-1}(A))^-$.

6.K. Check whether each of the following formulas holds for *every* metric space X and *every* subset Y of X.
  (a) $(X\backslash Y)^- = X\backslash Y^-$.
  (b) $(X\backslash Y)^\circ = X\backslash Y^\circ$.
  (c) $(X\backslash Y)^- = X\backslash Y^\circ$.
  (d) $(X\backslash Y)^\circ = X\backslash Y^-$.

6.L. Check whether each of the following formulas is true for *every* metric space X and *all* subsets Y and Z of X. (Give a proof or a counterexample.)
  (a) $(Y \cap Z)^- = Y^- \cap Z^-$.
  (b) $(Y \cup Z)^- = Y^- \cup Z^-$.
  (c) $(Y \cap Z)^\circ = Y^\circ \cap Z^\circ$.
  (d) $(Y \cup Z)^\circ = Y^\circ \cup Z^\circ$.

6.M. Let X be a metric space.
  For $Y, Z \subset X$, we say that Y is *dense in* Z if

$$Y \subset Z \subset Y^-.$$

A *dense* subset of X is simply a subset of X that is dense in X, i.e., a subset of X whose closure is X itself. For instance, as every element of $\mathbb{R}$ is adherent to $\mathbb{Q}$ (and $\mathbb{Q} \subset \mathbb{R}$), $\mathbb{Q}$ is dense in $\mathbb{R}$. (So is $\mathbb{R}\backslash\mathbb{Q}$; see the last lines of 6.1.)
  Prove that, for $Y \subset X$, *the following are equivalent*:
  (α) Y is dense in X.

(β) *Every element of X is the limit of a sequence of elements of Y.*

(γ) *Every nonempty open subset of X contains a point of Y.*

6.N.   (i) Let X be a metric space; let $Y, Z \subset X$ be such that Y is dense in Z and Z is dense in X. Prove that Y is dense in X.

(ii) Give an example of a metric space X and two dense subsets of X whose intersection is not dense in X.

(iii) Let X be a metric space and Y a dense subset of X. Does it follow that for every $A \subset X$, $Y \cap A$ is dense in A'?

7

CHAPTER

Completeness

7.1

We start with a problem from analysis.

If g is a continuous function on $[0, 1]$, then its restriction to $\mathbb{Q} \cap [0, 1]$ is continuous. Conversely, if f is a continuous function on $\mathbb{Q} \cap [0, 1]$, can one always extend it to a continuous function g on $[0, 1]$? In general, one cannot. Consider the function f defined by

$$f(x) := 0 \quad \text{if} \quad x \in \mathbb{Q} \cap [0, 1], \ x < 1/\sqrt{2},$$

$$f(x) := 1 \quad \text{if} \quad x \in \mathbb{Q} \cap [0, 1], \ x > 1/\sqrt{2}.$$

Owing to the irrationality of $1/\sqrt{2}$, this f is continuous at every point of its domain $\mathbb{Q} \cap [0, 1]$, but clearly it cannot be the restriction of a continuous function on $[0, 1]$.

Indeed, a continuous functions on $[0, 1]$ must be uniformly continuous. Then so is its restriction to $\mathbb{Q} \cap [0, 1]$. (Check this.) Thus, we arrive at a better question: If f is a uniformly continuous function on $\mathbb{Q} \cap [0, 1]$, can one always extend it to a continuous function g on $[0, 1]$? This time the answer turns out to be affirmative. We give the beginning of a proof; see Theorem 8.11 for further details.

Take a number a in $[0, 1]$ that does not lie in $\mathbb{Q} \cap [0, 1]$. What value should g have at a? There is a sequence $x_1, x_2, \ldots$ in $\mathbb{Q} \cap [0, 1]$ that tends to a. If a continuous extension g of f is at all possible, we must have

$$g(a) = \lim_{n \to \infty} g(x_n) = \lim_{n \to \infty} f(x_n).$$

It seems a reasonable strategy to use such a formula to define $g(a)$ for all $a \in [0, 1]$ and then to prove that the resulting function g is continuous.

But can we? We would have to know that the sequence

$$f(x_1), f(x_2), \ldots \qquad (*)$$

converges. It does, but that is not at all evident.

What we can see is that the terms of the sequence $(*)$ come close to each other. Indeed, as f is uniformly continuous, there exists a $\delta > 0$ such that

$$x, y \in \mathbb{Q} \cap [0, 1], \ |x-y| < \delta \implies |f(x)-f(y)| < 10^{-3}.$$

There is an N such that $|x_n - a| < \delta/2$ as soon as $n \geq N$. If now $n, m \geq N$, then $|x_n - x_m| < \delta$, so $|f(x_n) - f(x_m)| < 10^{-3}$.

In the same way, for any $\varepsilon > 0$ there is an N with

$$n, m \geq N \implies |f(x_n) - f(x_m)| < \varepsilon.$$

A sequence with this property is called a "Cauchy sequence" and we will see that every Cauchy sequence in $\mathbb{R}$ converges (Theorem 7.6). One encounters such sequences in many metric spaces, which is sufficient reason to consider them in a general setting.

7.2

Let (X, d) be a metric space. A sequence $(x_n)_{n \in \mathbb{N}}$ in X is called *Cauchy* if for every $\varepsilon > 0$, there is an N in $\mathbb{N}$ such that

$$n, m \geq N \implies d(x_n, x_m) < \varepsilon.$$

There are various equivalent definitions. For one of them we need a bit of preparation.

A nonempty subset Y of X is said to be *bounded* if the set $\{d(y, y') : y, y' \in Y\}$ has upper bounds in $\mathbb{R}$. (See 2.12.) Then we define the *diameter* of Y to be the number

$$\operatorname{diam} Y := \sup\{d(y, y') : y, y' \in Y\}.$$

By definition, the empty set is bounded and has diameter 0. Observe: If Y is bounded and $Z \subset Y$, then Z is also bounded, and $\operatorname{diam} Z \leq \operatorname{diam} Y$.

If $x_1, x_2, \ldots$ is a Cauchy sequence, then the set $\{x_1, x_2, \ldots\}$ is bounded. Indeed, there is a P in $\mathbb{N}$ such that $d(x_n, x_m) \leq 1$ as soon as $n, m \geq P$. Then,

$$d(x_n, x_P) \leq 1 + d(x_1, x_P) + \cdots + d(x_{P-1}, x_P)$$

for all $n \in \mathbb{N}$. The boundedness follows easily.

Lemma 7.3

For a sequence $(x_n)_{n \in \mathbb{N}}$ in a metric space (X, d), conditions (α), (β), and (γ) are equivalent.

(α) *The sequence is Cauchy.*
(β) $\lim_{n\to\infty} diam\{x_n, x_{n+1}, \ldots\} = 0.$
(γ) *There exist* $\varepsilon_1, \varepsilon_2, \ldots$ *in* $[0, \infty)$ *such that*

$$\left[\begin{array}{l} \lim_{n\to\infty} \varepsilon_n = 0, \\[2mm] d(x_n, x_m) \le \varepsilon_n \text{ for all } m \ge n. \end{array} \right.$$

Proof
(α) $\Longrightarrow$ (β) Obvious, once you know that the diameters exist.
(β) $\Longrightarrow$ (γ) Take $\varepsilon_n := diam\{x_n, x_{n+1}, \ldots\}$.
(γ) $\Longrightarrow$ (α) $d(x_n, x_m) \le 2\varepsilon_N$ if $n, m \ge N$. ∎

7.4
Every convergent sequence is Cauchy. [Let $x_n \to a$ and $\varepsilon > 0$. There is an N such that $d(x_n, a) < \varepsilon/2$ for all $n \ge N$. Then $d(x_n, x_m) < \varepsilon$ for $n, m \ge N$.]

Not every Cauchy sequence converges. In the metric space $(0, 1]$ (under the usual metric), the sequence $1, \frac{1}{2}, \frac{1}{3}, \ldots$ is Cauchy, but not convergent. It is true that $(0, 1]$ can be extended to a larger metric space in which the sequence does converge, but that is beside the point.

For another example, take $\mathbb{Z}$ with the 10-adic metric of Example 5.5(v). Consider the sequence

$$1, 11, 111, 1111, \ldots,$$

i.e., $x_1, x_2, \ldots$, where

$$x_n = 1 + 10 + 10^2 + \cdots + 10^{n-1}.$$

If $n, m \ge N$, then $x_n - x_m$ is divisible by 10^N, so $d(x_n, x_m) < N^{-1}$. Hence, the sequence is Cauchy. Can it converge? Suppose $a \in \mathbb{Z}$ and $x_n \to a$. Then [see Example 5.9(v)] $9x_n \to 9a$ and $9x_n+1 \to 9a+1$. But $9x_n+1 = 10^n \to 0$, so $9a+1 = 0$, which is impossible in $\mathbb{Z}$!

7.5
We call a metric space (X, d) *complete* if every Cauchy sequence in X converges (to a point of X). By the above, $(0, 1]$ with the ordinary metric and $\mathbb{Z}$ with the 10-adic metric are not complete. The standard example of a noncomplete space is $\mathbb{Q}$: For every irrational number a, the sequence

$$[a], \frac{[2a]}{2}, \frac{[3a]}{3}, \ldots$$

is Cauchy but (in $\mathbb{Q}$!) not convergent.
The prime example of a complete metric space is $\mathbb{R}$.

Theorem 7.6
$\mathbb{R}$ *and* $\mathbb{R}^N$ *(Euclidean metrics) are complete.*

Proof

(I) Let $x_1, x_2, \ldots$ be a Cauchy sequence in $\mathbb{R}$. As we have seen in 7.2, the sequence is bounded. Then by the Bolzano-Weierstrass Theorem it has a subsequence $x_{\alpha(1)}, x_{\alpha(2)}, \ldots$ that converges to some number a. Take $\varepsilon_1, \varepsilon_2, \ldots$ as in (γ) of Lemma 7.3. With Lemma 3.4 we obtain

$$d(x_n, a) \leq d(x_n, x_{\alpha(n)}) + d(x_{\alpha(n)}, a) \leq \varepsilon_n + d(x_{\alpha(n)}, a)$$

for all n. Hence, $x_n \to a$.

(II) Let $x_1, x_2, \ldots$ be a Cauchy sequence in $\mathbb{R}^N$. As earlier, we indicate by $(x_n)_i$ the ith coordinate of x_n. For each i, we have

$$|(x_n)_i - (x_m)_i| \leq \|x_n - x_m\| \qquad (n, m \in \mathbb{N})$$

so that the sequence $(x_1)_i, (x_2)_i, \ldots$ is Cauchy in $\mathbb{R}$. Let a_i be its limit. Then $a := (a_1, \ldots, a_N)$ is an element of $\mathbb{R}^N$, and $x_n \to a$ since $(x_n)_i \to a_i$ for each i. ∎

Theorem 7.7

A closed subset of a complete metric space is complete under the induced metric. [See 5.4(v).]

Proof

Let X be a complete metric space and Y a closed subset of X. If $x_1, x_2, \ldots$ is a Cauchy sequence in Y, then it is also a Cauchy sequence in X. Hence, there is an a in X with $x_n \to a$. But Y is closed, so $a \in Y$. ∎

7.8

A subset A of a metric space X is called *dense* in X if every nonempty open subset of X contains a point of A. (This is in accordance with Exercise 6.M.)

Since in $\mathbb{R}$ every nonempty open subset contains an open interval, we see that a subset of $\mathbb{R}$ is dense if it intersects every interval. For instance, $\mathbb{Q}$ and $\mathbb{R} \backslash \mathbb{Q}$ are dense in $\mathbb{R}$.

The following theorems express the fact that certain sets are "large." In what sense this is meant we will discuss after the theorems.

Theorem 7.9 (Baire Category Theorem; first version)

If X is a complete metric space and $U_1, U_2, \ldots$ are open and dense subsets of X, then their intersection is dense in X (although possibly not open).

Proof

Take an open set V which is nonempty. Our proof produces an element a in $V \cap (\bigcap_{n=1}^{\infty} U_n)$.

Take a_1 in the set $V \cap U_1$, which is nonempty because U_1 is dense in X. Since $V \cap U_1$ is open, there is an $\varepsilon_1 > 0$ such that the ball $B_{2\varepsilon_1}(a_1)$ is contained in $V \cap U_1$. Then, by Exercise 6.I

$$B_{\varepsilon_1}(a_1)^- \subset V \cap U_1.$$

Next, take $a_2 \in B_{\varepsilon_1}(a_1) \cap U_2$ and $\varepsilon_2 > 0$ with $B_{2\varepsilon_2}(a_2) \subset B_{\varepsilon_1}(a_1) \cap U_2$; then,

$$B_{\varepsilon_2}(a_2)^- \subset B_{\varepsilon_1}(a_1) \cap U_2.$$

Continuing in this way, we get a sequence $a_1, a_2, \ldots$ in X and a sequence $\varepsilon_1, \varepsilon_2, \ldots$ in $\mathbb{R}$ with

$$B_{\varepsilon_n}(a_n)^- \subset B_{\varepsilon_{n-1}}(a_{n-1}) \cap U_n \subset V \cap U_1 \cap \cdots \cap U_n$$

for each n. By making ε_n smaller if needed, we may assume that $\varepsilon_n < n^{-1}$. Then,

$$d(a_n, a_m) \le n^{-1} \text{ for all } m \ge n,$$

because $a_m \in B_{\varepsilon_n}(a_n) \subset B_{1/n}(a_n)$. Thus, $a_1, a_2, \ldots$ is a Cauchy sequence and converges to some $a \in X$; but also

$$a \in \{a_n, a_{n+1}, \ldots\}^- \subset B_{\varepsilon_n}(a_n)^- \subset V \cap U_1 \cap \ldots \cap U_n$$

for all n, so $a \in V \cap \left(\bigcap_{n=1}^{\infty} U_n\right)$. ∎

The following result is a reformulation of the above theorem.

Theorem 7.10
If a complete metric space is written as a union of countably many closed sets, then at least one of those closed sets contains a ball.

Proof
Let $X = \bigcup_{n=1}^{\infty} A_n$ with A_n closed in X for all n. Then each $X \backslash A_n$ is open, and $\bigcap_{n=1}^{\infty} X \backslash A_n = X \backslash \bigcup_{n=1}^{\infty} A_n = \varnothing$. Thus, not every $X \backslash A_n$ can be dense, say, $X \backslash A_k$ is not dense. Then, for some nonempty open set V, we have $V \cap (X \backslash A_k) = \varnothing$, that is, $V \subset A_k$; but then A_k contains a ball. ∎

7.11
Either of the above theorems is referred to as the *Baire Category Theorem*. To explain where the word "category" comes from, we need the following. A subset A of a metric space X is said to be *nowhere dense in* X if X has no point interior to the closure of A; that is, A^- contains no ball. For instance, $\mathbb{N}$ and the Cantor set are nowhere dense in $\mathbb{R}$. The complement of a dense open set is a closed nowhere dense set. A subset of X is called *meager in* X if it is a union of countably many sets that are nowhere dense in X. Every nowhere dense set is, of course, meager; but also $\mathbb{Q}$, which is not nowhere dense in $\mathbb{R}$, is meager in $\mathbb{R}$.

The idea behind this definition is that nowhere dense sets are rather small, and so are unions of "not too many" nowhere dense sets. The following three observations may serve to illustrate this:

(a) Every subset of a meager set is meager.
(b) Every union of countably many meager sets is meager.
(c) A complete metric space is not meager. (This is Theorem 7.10.)

Meager subsets of a metric space X are also called *sets of the first category in X*; all other subsets of X are *sets of the second category*. As you notice, these terms are not particularly descriptive; that is why we prefer the word "meager."

Be careful, "meager," like "closed" but unlike "complete," is a relative term. Properly speaking, a set cannot be just "meager"; it can be "meager in" a given metric space. The Cantor set is meager in $\mathbb{R}$, not in itself. Statement (c), above, really is a bit careless.

Examples 7.12
We discuss two examples of how to apply the Baire Category Theorem.

(i) $\mathbb{Q}$ *is not the intersection of countably many open subsets of* $\mathbb{R}$. Indeed, $\mathbb{Q}$ can be written as $\{q_1, q_2, \ldots\}$. If $\mathbb{Q}$ were equal to $\bigcap_{n=1}^{\infty} U_n$ where each U_n is open in $\mathbb{R}$, then $\bigcap_{n=1}^{\infty} U_n \backslash \{q_n\} = \varnothing$. However, $\mathbb{Q} \subset U_n$ for all $n \in \mathbb{N}$ and then all U_n are dense. Then, all $U_n \backslash \{q_n\}$ are open and dense (?), and according to the Baire Category Theorem their intersection is nonempty (even dense).

(ii) The indicator function of $\mathbb{Q}$ is not continuous on $\mathbb{R}$. Somewhat more can be said using (i), above: *There does not exist a sequence* $f_1, f_2, \ldots$ *of continuous functions on* $\mathbb{R}$ *such that* $f_n(x) \to \mathbf{1}_{\mathbb{Q}}(x)$ *for all* $x \in \mathbb{R}$. Indeed, if $f_1, f_2, \ldots$ is such a sequence and U_m is the inverse image of $(\frac{1}{2}, \infty)$ under f_m, then $\bigcup_{m \geq n} U_m$ is open for each n, because f_m is continuous. However,

$$\bigcap_{n=1}^{\infty} \bigcup_{m \geq n} U_m = \mathbb{Q}$$

since $f_n(x) \to \mathbf{1}_{\mathbb{Q}}(x)$ for all $x \in \mathbb{R}$. But according to (i), this is impossible.

A more interesting (and harder) application of the Baire Category Theorem is Exercise 8.I.

7.13
For a metric space (X, d), a map $f : X \to X$ is called a *contraction* if it is a Lipschitz map with Lipschitz constant strictly less than 1, i.e., if there is a $K < 1$ with

$$d(f(x), f(y)) \leq K d(x, y) \text{ for all } x, y \in X.$$

[See Example 5.9(ii).] As another application of completeness we have the following theorem.

Theorem 7.14 (Banach's Contraction Principle or Banach's Fixed Point Theorem)

If (X,d) is a nonempty complete metric space and $f : X \to X$ is a contraction, then there exists a unique $x \in X$ with $f(x) = x$ (i.e., f has a unique fixed point).

Proof

Let $K < 1$ be a Lipschitz constant for f. Take $x_1 \in X$ (any x_1 will do) and define a sequence $(x_n)_{n \in \mathbb{N}}$ by

$$x_{n+1} := f(x_n). \tag{1}$$

Then, we have for every $p > 1$ that

$$d(x_{p+1}, x_p) = d\big(f(x_p), f(x_{p-1})\big) \le K d(x_p, x_{p-1}).$$

It follows that for all $p \in \mathbb{N}$,

$$d(x_{p+1}, x_p) \le K^{p-1} d(x_2, x_1). \tag{2}$$

Then, for all $m < n$

$$
\begin{aligned}
d(x_n, x_m) &\le d(x_n, x_{n-1}) + d(x_{n-1}, x_{n-2}) + \cdots + d(x_{m+1}, x_m) \\
&\le (K^{n-2} + K^{n-3} + \cdots + K^{m-1}) d(x_2, x_1) \\
&\le \left(\sum_{l=m-1}^{\infty} K^l \right) d(x_2, x_1) \\
&= \frac{K^{m-1}}{1-K} d(x_2, x_1).
\end{aligned}
$$

Thus, $(x_n)_{n \in \mathbb{N}}$ is a Cauchy sequence. By the completeness of X, there exists an $x \in X$ such that $x_n \to x$. For every n, we obtain from (1) and (2)

$$
\begin{aligned}
d\big(f(x), x\big) &\le d\big(f(x), f(x_n)\big) + d(x_{n+1}, x_n) + d(x_n, x) \\
&\le K d(x, x_n) + K^{n-1} d(x_2, x_1) + d(x_n, x).
\end{aligned}
$$

As $d(x, x_n) \to 0$ and $K^{n-1} \to 0$, we have $d\big(f(x), x\big) = 0$, and $f(x) = x$.

For the uniqueness, note that if $f(x) = x$ and $f(y) = y$, then

$$d(x, y) = d\big(f(x), f(y)\big) \le K d(x, y)$$

and, thus, $d(x, y) = 0$ and $x = y$. ∎

Example 7.15

Let $K < 1$ and let $f : \mathbb{R} \to \mathbb{R}$ be a differentiable function for which

$$|f'(x)| \le K \text{ for all } x \in \mathbb{R}.$$

It then follows from the Mean Value Theorem that f is a contraction. Thus, the equation

$$f(x) = x$$

has one solution in $\mathbb{R}$. Moreover, as we can see from the above proof, for every $a \in \mathbb{R}$, the sequence

$$a, f(a), f(f(a)), f(f(f(a))), \ldots$$

converges to this solution.

For a remarkable application of the Contraction Principle to a differential equation, see 8.9.

Extra: Meager Sets and the Mazur Game

In 7.11 we have observed that for subsets of a complete metric space X, meagerness is a kind of smallness, as is shown by the following properties:

(a) Subsets of meager sets are meager.
(b) Unions of countably many meager sets are meager.
(c) The entire space X is not meager.

It is sometimes convenient to say that "most" elements of X have a certain property if the elements that do not have the property form a meager set. For example, $\mathbb{Q}$ is a meager subset of $\mathbb{R}$, so "most" real numbers are irrational.

Meager sets turn up in many places in Topology and Analysis. For instance, as you may know, the derivative of a differentiable function on $\mathbb{R}$ need not be continuous everywhere; but it can be shown to be continuous at "most" points of $\mathbb{R}$. For another example, in Exercise 3.H we considered separately continuous functions on $\mathbb{R}^2$ and we saw that such a function might not be continuous everywhere. However, a separately continuous function has to be continuous at "most" points of $\mathbb{R}^2$.

Meager sets are also valuable tools for certain proofs and constructions—for instance, to make continuous functions on $[0, 1]$ that are differentiable at *no* point of $[0, 1]$. What actually happens there is that one considers the set of all continuous functions on $[0, 1]$, puts a suitable metric on it, and shows that "most" continuous functions are nowhere differentiable. (In Exercise 8.I we apply this technique to obtain a related result. If you are interested in continuous nowhere differentiable functions, see 8.10.)

For a very different situation in which meager sets come up, we need an excursion into the theory of games.

Consider this game for two players, Red and Black. Nine counters are placed on a table. A "move" consists of taking away one or two counters. Starting with Red, the players alternately make such moves until all counters are gone. The player who takes the last one wins the game. Who would you prefer to be, Red (the first player) or Black?

No matter how smart Red is, Black can always beat him by following this recipe: If Red in his first move takes x counters, Black takes $3 - x$; that is allowable and leaves six counters on the table. If Red now takes y of them, Black takes $3 - y$, leaving three counters. Red takes z, Black the remaining $3 - z$, and wins. You see that Red is defenseless: Black has a "winning strategy."

The same game with ten counters instead of nine has a different outcome. This time, Red can open by taking one counter, thereby reducing the game to the previous one with the crucial difference that now their roles are reversed. By adopting the second player's strategy for the nine-counter game, Red is sure to win. In the ten-counter game the first player has a winning strategy.

In 1928 the Polish mathematician Stanislaw Mazur proposed a certain collection of infinite games, one for each subset of $\mathbb{R}$. The "Mazur game" for a set A runs as follows. (For convenience, we present a slight modification of the original version.) There are two players, Red and Black. Red opens the game by choosing a closed interval I_1 of length at most 1; Black chooses a closed interval I_2 of length at most $1/2$, contained in I_1; Red chooses a closed interval $I_3 \subset I_2$ of length at most $1/3$; and so on. Together, they build a sequence

$$I_1 \supset I_2 \supset I_3 \supset \cdots$$

of intervals. By the Cantor Theorem (2.19) the intersection of all these intervals contains precisely one number. Red wins if this number belongs to the set A, Black if it does not.

Question: Does either player have a winning strategy?

If, say, A is $[0, 1]$, Red has an easy job; by choosing $I_1 = [0, 1]$ his victory is certain. If A is $\mathbb{Q}$, Black can always win by doing the following. The elements of $\mathbb{Q}$ can be put into a sequence $q_1, q_2, q_3, \ldots$. Now after Red has selected I_1, Black chooses I_2 so that it does not contain q_1. In his next turn, he makes sure that $q_2 \notin I_4$, and so on. The number in the intersection of all intervals I_n cannot be q_1 or q_2 or $\ldots$; in short, it lies outside $\mathbb{Q}$ and Black wins.

For many sets, it is unknown if Red or Black has a strategy, but there is one fairly general result that brings us back to meager sets: *If A is meager, Black has a winning strategy; if $\mathbb{R}\backslash A$ is meager, Red has one.* Let us prove the first part.

Suppose A is meager. A is contained in the union of closed sets $A_1, A_2, \ldots$, each with empty interior. Red selects his first interval, $I_1 = [a, b]$, say. As A_1 has empty interior, the open interval (a, b) cannot be entirely contained in A_1, so $(a, b)\backslash A_1$ is a nonempty open set. Now Black can choose I_2 so that $I_2 \subset (a, b)\backslash A_1$; then, in particular, $I_2 \cap A_1 = \varnothing$.

In the next round, he can make sure that $I_4 \cap A_2 = \emptyset$, and so on. The resulting number cannot lie in any A_n and therefore is not an element of A. You see that this procedure gives Black a winning strategy.

You can verify for yourself that Red has a strategy if $\mathbb{R} \backslash A$ is meager.

Whatever set A is, it is clearly impossible for both players to have winning strategies. You might expect that in any case either Red or Black has one, but it is not as simple as that. In our Extra in Chapter 17 we discuss a set-theoretic assumption, the Axiom of Choice. If you accept this axiom as being true (and most mathematicians do), you can prove that there exist sets A for which neither player has a strategy. If you reject the Axiom of Choice (a point of view that is respectable, if not popular), you might prefer the "Axiom of Determinateness" that says that for every Mazur game, one player has a strategy.

In mathematics there is nothing frivolous about games and strategies. An extensive and growing part of mathematics is "game theory." It was started in 1944 by John von Neumann and Oskar Morgenstern. Dealing almost exclusively with *finite* games, it has wide applications, e.g., in economics, sociology, and politics. (In 1994 the Nobel prize for economics was awarded to three game theorists.) The terminology of *infinite* games is occasionally very enlightening in set theory.

Further reading

Oxtoby, J.C., *Measure and Category*, Springer-Verlag, New York, 1970.

Exercises

7.A. We know that the Euclidean metric and the $\tan^{-1}$ metric on $\mathbb{R}$ determine the same convergent sequences. [See Example 5.5(ii).] Show that they do not determine the same Cauchy sequences and that, in fact, $\mathbb{R}$ is *not* complete relative to the $\tan^{-1}$ metric.

7.B. Let X be the unit sphere in $\mathbb{R}^3$:

$$X := \{x \in \mathbb{R}^3 : \|x\| = 1\}.$$

On X we consider the Euclidean metric d_E and the "surface-metric" d:

$$d(x, y) := \cos^{-1}\langle x, y \rangle \qquad (x, y \in X).$$

[see Example 5.3(iv).]

   Show that d_E and d define the same convergent sequences.

   Show that they also define the same Cauchy sequences.

   Show that (X, d) is complete.

7.C. For $x, y \in \mathbb{N}$, set $\delta(x, y) := |x-y|/xy$.
   (i) Show that δ is a metric, similar (see 6.20) to the Euclidean metric on $\mathbb{N}$.
   (ii) Show that $(n)_{n\in\mathbb{N}}$ is a Cauchy sequence for δ but not a Cauchy sequence for the Euclidean metric d_E.
   (iii) Show that $(\mathbb{N}, d_E)$ is complete, whereas $(\mathbb{N}, \delta)$ is not.

7.D. Equip $[0, 1)$ with the metric

$$d(x, y) := \left| \frac{x}{1-x} - \frac{y}{1-y} \right|.$$

Is the metric space $([0, 1), d)$ complete?

7.E. Let Y be a subset of a metric space (X, d) and let d_Y be the induced metric in Y. [See 5.4(v).]
   Prove: If the metric space (Y, d_Y) is complete, then Y is d-closed in X.
   (Compare with Theorem 7.7. Apparently, for subsets of a complete metric space, completeness and closedness are synonymous.)

7.F. (i) Consider the linear system of equations

$$\left.\begin{array}{ccccc} a_{11}x_1 & +\cdots+ & a_{1N}x_N & = & b_1 \\ \vdots & & \vdots & & \vdots \\ a_{N1}x_1 & +\cdots+ & a_{NN}x_N & = & b_N \end{array}\right] \tag{$*$}$$

   Writing

$$c_{ii} = 1 - a_{ii} \quad \text{for} \quad i \in \{1, \ldots, N\},$$

$$c_{ik} = -a_{ik} \quad \text{for} \quad i, k \in \{1, \ldots, N\}, \; i \neq k,$$

   we see that $(*)$ can be rewritten as

$$x_i = \sum_{k=1}^{N} c_{ik}x_k + b_i \qquad (i = 1, \ldots, N). \tag{$**$}$$

   Show that $(*)$ has a unique solution in case

$$\sum_{i,k} c_{ik}^2 < 1.$$

   (Apply the Contraction Principle to a suitable map $f : \mathbb{R}^N \to \mathbb{R}^N$. Use the Cauchy-Schwarz Inequality to prove that f is a contraction.)
   (ii) Prove the same result (with the same condition on the c_{ik}) for the nonlinear system of equations

$$x_i = \sum_{k=1}^{N} \sin(c_{ik}x_k) + b_i \qquad (i = 1, \ldots, N).$$

7.G. A point a in a metric space X is called *isolated* if $\{a\}$ is an open set. Thus, $\mathbb{R}$ and $\mathbb{Q}$ have no isolated points; in $\mathbb{N}$, every point is isolated.
   Let X be a nonempty complete metric space and assume X is countable:

$$X = \{x_1, x_2, \ldots\}.$$

For every $n \in \mathbb{N}$, define $U_n := X \backslash \{x_n\}$. The sets $U_1, U_2, \ldots$ are open (why?) and have empty intersection. Deduce from the Baire Category Theorem that some $\{x_n\}$ has to be open.

Thus: *Every nonempty metric space that is complete and countable has an isolated point.*

7.H. Let X be a metric space, $A \subset B \subset X$.

  (i) If B is meager in X, must A be meager in X?

  (ii) If A is meager in B, must A be meager in X?

  (iii) If A is meager in X, must A be meager in B?

  (iv)–(vi) As for (i)–(iii), but with "nonmeager" instead of "meager."

8

CHAPTER

Uniform
Convergence

8.1

In analysis, one often has to do with a sequence $f_1, f_2, \ldots$ of functions converging to a function g, and the question arises of whether the integrals of the f_n over some given interval $[a, b]$ converge to the integral of g.

Consider the functions f_n on $[0, 1]$ defined by

$$f_n(t) := nt^n(1-t^n) \qquad (t \in [0, 1]).$$

We have $f_n(t) \to 0$ for every t, but *not* $\int_0^1 f_n(t)dt \to 0$. Indeed,

$$\int_0^1 f_n(t)dt = n \int_0^1 (t^n - t^{2n})dt$$

$$= n\left(\frac{1}{n+1} - \frac{1}{2n+1}\right) = \frac{n^2}{(n+1)(2n+1)} \to \frac{1}{2}.$$

At a first glance, this phenomenon seems strange, but a look at the graphs of the functions (see next page) is enlightening.

A sequence $f_1, f_2, \ldots$ of functions on a set S is said to *converge pointwise* to a function g if $f_n(s) \to g(s)$ for every $s \in S$.

Apparently, pointwise convergence does not entail convergence of the integrals. In the present section we consider a metric d_∞ on the set of all continuous functions on a given interval such that d_∞-convergence is better behaved relative to integration.

It pays to take a more general point of view and consider functions defined on an arbitrary nonempty set, not necessarily an interval or not

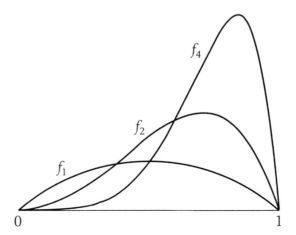

even a subset of $\mathbb{R}$ at all. We will see then how abstract topological methods can be applied to problems in advanced analysis (e.g., differential equations).

Throughout this section S is a nonempty set.

8.2

A function f on S is *bounded* if the set of its values is bounded, i.e. if there exists a number M such that

$$|f(s)| \le M \quad \text{for all } s \in S. \tag{$*$}$$

By $l^\infty(S)$ we denote the set of all bounded functions on S.

Take $f \in l^\infty(S)$. Formula $(*)$ is equivalent to saying that M is an upper bound of the set $\{|f(s)| : s \in S\}$. By the Least Upper Bound Theorem, among those upper bounds there is a smallest one. That number we call $\|f\|_\infty$:

$$\|f\|_\infty \text{ is the smallest } M \text{ satisfying } (*),$$

or

$$\|f\|_\infty := \sup\{|f(s)| : s \in S\}.$$

Lemma 8.3

Let $f, g \in l^\infty(S)$. Then $f + g$ [the function $s \mapsto f(s) + g(s)$] belongs to $l^\infty(S)$ and

$$\|f + g\|_\infty \le \|f\|_\infty + \|g\|_\infty.$$

Proof

Set $M := \|f\|_\infty$ and $N := \|g\|_\infty$. For all $s \in S$ we have

$$|(f + g)(s)| = |f(s) + g(s)| \le |f(s)| + |g(s)| \le M + N.$$

The lemma follows.   ∎

8.4

Similarly, if $f, g \in l^\infty(S)$, then $f - g \in l^\infty(S)$. This observation enables us to define

$$d_\infty(f, g) := \|f - g\|_\infty \qquad (f, g \in l^\infty(S)).$$

(The resemblance with the Euclidean metric on $\mathbb{R}^N$ is evident.) One easily verifies that d_∞ is a metric.

For all $f, g \in l^\infty(S)$ and $s \in S$, we have

$$|f(s) - g(s)| \le d_\infty(f, g).$$

Hence, d_∞-convergence implies pointwise convergence. The converse is false, as can be seen from the example at the beginning of this section. There, we had the functions $f_n : t \mapsto nt^n(1 - t^n)$ on the set $[0, 1]$. The sequence converges to the zero function pointwise. But for every n,

$$d_\infty(f_n, 0) = \|f_n\|_\infty = \sup\{|f_n(t)| : t \in [0, 1]\}.$$

As $f_n(t) \ge 0$ for every t, $\|f_n\|_\infty$ is simply the largest value of f_n on $[0, 1]$, which by the usual calculus techniques is seen to be $n/4$. Thus, $d_\infty(f_n, 0) \to \infty$!

To get an insight in the difference between pointwise convergence and d_∞-convergence, imagine we are given functions $g, f_1, f_2, \ldots$ on S. If we wish to show that $f_n \to g$ in the sense of d_∞, then, given a positive ε, we have to find an N such that for all $n \ge N$,

$$\|f_n - g\|_\infty \le \varepsilon,$$

i.e.,

$$|f_n(s) - g(s)| \le \varepsilon \quad \text{for all } s.$$

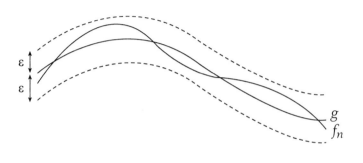

On the other hand, if we want to prove pointwise convergence, then we are confronted with an s and an ε, and we need an N such that for all $n \geq N$,

$$|f_n(s) - g(s)| \leq \varepsilon.$$

For pointwise convergence, N is allowed to depend on s; d_∞-convergence requires an N that does the job for all s at a time and that is much harder. The difference between pointwise and d_∞-convergence is not unlike that between ordinary and uniform continuity: See 3.11.

For functions $g, f_1, f_2, \ldots$ on S, we say that

$$f_n \to g \quad uniformly \text{ on } S$$

or also that

$$f_n(s) \to g(s) \quad uniformly \text{ for } s \in S$$

if $\|f_n - g\|_\infty \to 0$. For bounded functions, this is precisely d_∞-convergence, but uniform convergence does not presuppose boundedness. Thus, we can say that

$$\sqrt{s^2 + n^{-1}} \to |s| \quad uniformly \text{ for } s \in \mathbb{R}$$

because $0 \leq \sqrt{s^2 + n^{-1}} - |s| \leq n^{-\frac{1}{2}}$ for all s and n.

The feedback to the opening of this section is easy to give:

Theorem 8.5

Let $[a,b]$ be an interval, $g, f_1, f_2, \ldots$ continuous functions on $[a,b]$ such that $f_n \to g$ uniformly. Then,

$$\int_a^b f_n(t)dt \to \int_a^b g(t)dt.$$

Proof

As $|f_n(t) - g(t)| \leq d_\infty(f_n, g)$ for all t, we have

$$\left| \int_a^b f_n(t)dt - \int_a^b g(t)dt \right| \leq (b-a)d_\infty(f_n, g). \qquad \blacksquare$$

What we have here is really a precise rendering of the phenomenon we vaguely described in 5.1: Integration is a continuous function $X \to \mathbb{R}$ if by X we indicate the set of all continuous functions on $[a, b]$ endowed with the metric d_∞.

Another application of uniform convergence:

Theorem 8.6

Let (X,d) be a metric space and let $f_1, f_2, \ldots$ be a sequence of functions on X, converging uniformly to a function g. Suppose every f_n is continuous. Then, so is g.

Proof

Take $a \in X$, $\varepsilon > 0$; we want to find a $\delta > 0$ such that

$$|g(x)-g(a)| < \varepsilon \quad \text{for all} \quad x \in B_\delta(a).$$

As $\|f_n-g\|_\infty \to 0$, there certainly exists a $P \in \mathbb{N}$ with

$$\|f_P-g\|_\infty < \frac{\varepsilon}{3}.$$

As f_P is continuous at a, there is a $\delta > 0$ such that

$$|f_P(x)-f_P(a)| < \frac{\varepsilon}{3} \quad \text{for all} \quad x \in B_\delta(a).$$

This δ does the trick: For $x \in B_\delta(a)$, we have

$$|g(x)-g(a)| \le |g(x)-f_P(x)| + |f_P(x)-f_P(a)| + |f_P(a)-g(a)|$$
$$\le \|g-f_P\|_\infty + |f_P(x)-f_P(a)| + \|f_P-g\|_\infty$$
$$< \frac{\varepsilon}{3} + \frac{\varepsilon}{3} + \frac{\varepsilon}{3} = \varepsilon. \qquad \blacksquare$$

Observe that the *pointwise* limit of the continuous functions $f_n : t \mapsto t^n$ ($t \in [0, 1]$) fails to be continuous. Observe also that even uniform convergence does not preserve differentiability: At the end of 8.4 we have seen that the nondifferentiable function $s \mapsto |s|$ is a uniform limit of differentiable ones.

For further applications of uniform convergence, we need:

Theorem 8.7

Under d_∞, $l^\infty(S)$ is a complete metric space.

Proof

Let $f_1, f_2, \ldots$ be a Cauchy sequence in $l^\infty(S)$:

$$\|f_m-f_n\|_\infty \le \varepsilon_n \quad \text{for all} \quad m \ge n,$$

where $\varepsilon_n \to 0$.

Take $s \in S$. If $m, n \in \mathbb{N}$ and $m \ge n$, then

$$|f_m(s)-f_n(s)| \le \|f_m-f_n\|_\infty \le \varepsilon_n. \tag{1}$$

It follows that $f_1(s), f_2(s), \ldots$ is a Cauchy sequence in $\mathbb{R}$. Let its limit be $g(s)$.

Thus, we obtain a function g on S. If in (1) we let m tend to infinity, we see that

$$|g(s)-f_n(s)| \le \varepsilon_n \tag{2}$$

for all n and s. Upon taking $n = 1$, we find $|g(s)| \le |f_1(s)|+\varepsilon_1 \le \|f_1\|_\infty +\varepsilon_1$, so that $g \in l^\infty(S)$. Then, (2) implies $\|g-f_n\|_\infty \le \varepsilon_n$ for all n, so $f_n \to g$ in the sense of d_∞. $\qquad \blacksquare$

For a closed bounded interval $[a, b]$, we denote by

$$C[a, b]$$

the set of all continuous functions on $[a, b]$.

Theorem 8.8
Under d_∞, $C[a,b]$ is a complete metric space.

Proof
We know (Corollary 3.10) that $C[a, b]$ is contained in $l^\infty([a, b])$, so d_∞ is in fact a metric on $C[a, b]$. It follows from Theorem 8.6 that $C[a, b]$ is (sequentially) closed in $l^\infty([a, b])$. Then, by the completeness of $l^\infty([a, b])$, $C[a, b]$ is complete, too. (Theorem 7.7.)   ∎

8.9
Let us see how this can be applied to differential equations.

We will show that there exists one and only one function h on $[0, 1]$ satisfying

$$\begin{bmatrix} h'(t) = \cos th(t) \quad (t \in [0, 1]), \\ h(0) = 0. \end{bmatrix} \tag{$*$}$$

If h satisfies $(*)$, then h is continuous, and for all $s \in [0, 1]$, we have $h(s) = h(0) + \int_0^s h'(t)dt$. Thus, $(*)$ implies

$$\begin{bmatrix} h \in C[0, 1], \\ h(s) = \int_0^s \cos th(t)dt \quad (s \in [0, 1]). \end{bmatrix} \tag{$**$}$$

Conversely, it is perfectly easy to see that every function h for which $(**)$ holds is a solution of $(*)$. Thus, we are done if we can prove that there exists precisely one h satisfying $(**)$.

For every $f \in C[0, 1]$, we define a function $f^\sim$ on $[0, 1]$ by

$$f^\sim(s) := \int_0^s \cos tf(t)dt \quad (s \in [0, 1]).$$

Then, $f^\sim$ is continuous. Hence,

$$f \mapsto f^\sim \quad (f \in C[0, 1]) \tag{1}$$

is a map $C[0, 1] \to C[0, 1]$. What we have to prove is that there exists precisely one h in $C[0, 1]$ for which $h = h^\sim$. In other words, we want to show that the map defined by (1) has precisely one fixed point.

Now $C[0, 1]$ is a complete metric space under d_∞. Hence, by the Banach Contraction Principle (7.14), we are done if the map (1) happens to be a contraction. We proceed to show that it is, and that actually

$$d_\infty(f^\sim, g^\sim) \le \tfrac{1}{2} d_\infty(f, g)$$

for all f and g in $C[0, 1]$.

Take $f, g \in C[0, 1]$; let $K := d_\infty(f, g)$. Then

$$|f(t) - g(t)| \le K \qquad (t \in [0, 1])$$

and we wish to show that

$$|f^\sim(s) - g^\sim(s)| \le \tfrac{1}{2} K \qquad (s \in [0, 1]).$$

We need the inequality

$$|\cos u - \cos v| \le |u - v| \qquad (u, v \in \mathbb{R}),$$

(following from the Mean Value Theorem, applied to $\dfrac{\cos u - \cos v}{u - v}$).

For every s in $[0, 1]$, we have

$$|f^\sim(s) - g^\sim(s)| = \left| \int_0^s \cos tf(t)dt - \int_0^s \cos tg(t)dt \right|$$

$$= \left| \int_0^s \big(\cos tf(t) - \cos tg(t)\big)dt \right|$$

$$\le \int_0^s \big|\cos tf(t) - \cos tg(t)\big| dt$$

$$\le \int_0^s \big|tf(t) - tg(t)\big| dt$$

$$\le \int_0^s tK\, dt = \tfrac{1}{2} Ks^2 \le \tfrac{1}{2} K,$$

and we are done.

8.10

As another application of the Contraction Principle, we prove the existence of a continuous function on $[0, 1]$ that is nowhere differentiable.

For X, we take the set of all continuous functions $f : [0, 1] \to [0, 1]$ with $f(0) = 0$, $f(1) = 1$. For every $f \in X$, there exists one $f^\sim : [0, 1] \to \mathbb{R}$ with

$$f^\sim(t) := \tfrac{3}{4} f(3t) \qquad\qquad \text{if} \ \ 0 \le t \le \tfrac{1}{3},$$

$$f^\sim(t) := \tfrac{1}{4} + \tfrac{1}{2} f(2 - 3t) \qquad \text{if} \ \ \tfrac{1}{3} \le t \le \tfrac{2}{3},$$

$$f^\sim(t) := \tfrac{1}{4} + \tfrac{3}{4} f(3t - 2) \qquad \text{if} \ \ \tfrac{2}{3} \le t \le 1.$$

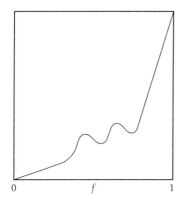

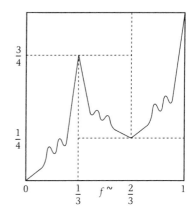

Then, $f^\sim(0) = 0, f^\sim(\frac{1}{3}) = \frac{3}{4}, f^\sim(\frac{2}{3}) = \frac{1}{4}$, and $f^\sim(1) = 1$. We see that $f^\sim$ is continuous and is, in fact, an element of X.

If $f, g \in X$ and if M is a number such that $|f(t)-g(t)| \le M$ for all t, then, clearly, $|f^\sim(t)-g^\sim(t)| \le \frac{3}{4} M$ for all t. Thus,

$$d_\infty(f^\sim, g^\sim) \le \tfrac{3}{4} d_\infty(f, g) \qquad (f, g \in X).$$

Therefore, the map $f \mapsto f^\sim$ is a contraction of X. Now, as uniform convergence implies pointwise convergence, X is a (sequentially) d_∞-closed subset of $C[0, 1]$. Then, X is a complete metric space. By the Banach Contraction Principle, there is an $h \in X$ with $h = h^\sim$.

Certainly, h is continuous. We prove that h is nowhere differentiable. To this end, we observe that for all $n \in \mathbb{N}$,

$$\left| h\left(\frac{k-1}{3^n}\right) - h\left(\frac{k}{3^n}\right) \right| \ge 2^{-n} \qquad (k = 1, 2, \ldots, 3^n). \qquad (*)$$

A proof is easily given by induction. The crucial observation is that for each n and for $k \in \{1, 2, \ldots, 3^n\}$, the numbers $(k-1)3^{-n}$ and $k3^{-n}$ both lie in $[0, \frac{1}{3}]$ or in $[\frac{1}{3}, \frac{2}{3}]$ or in $[\frac{2}{3}, 1]$, according to whether $k \le 3^n$ or $3^n < k \le 2.3^n$ or $2.3^n < k$.

Now take $a \in [0, 1]$. For $n \in \mathbb{N}$, there exists a k in $\{1, 2, \ldots, 3^n\}$ with $(k-1)3^{-n} \le a \le k3^{-n}$. It follows from $(*)$ that the interval $[(k-1)3^{-n}, k3^{-n}]$ contains a number t_n (e.g., one of the end points) such that

$$|h(t_n)-h(a)| \ge 2^{-n-1}.$$

But, $|t_n-a| \le 3^{-n}$, so

$$\lim_{n\to\infty} \left| \frac{h(t_n)-h(a)}{t_n-a} \right| = \infty$$

and h cannot be differentiable at a.

Look back at the problem posed at the beginning of Section 7: If f : $\mathbb{Q} \cap [0, 1] \to \mathbb{R}$ is uniformly continuous, can one extend it to a continuous function on $[0, 1]$? We now have enough tools to show that the strategy we have outlined really works.

Theorem 8.11

Let Y be a dense subset of a metric space X and let f be a uniformly continuous map of Y into a complete metric space X'.

(i) *There is a unique continuous $g: X \to X'$ whose restriction to Y is f. This g is uniformly continuous.*
(ii) *If f is Lipschitz, so is g. If f is an isometry, so is g.*

Proof

(i) Let d and d' be the metrics of X and X'. For simplicity of notation, write $y' := f(y)$ if $y \in Y$. Let us say that an element a of X is associated with a b in X' if there exist $y_1, y_2, \ldots$ in Y with

$$y_n \to a \quad \text{and} \quad y'_n \to b.$$

Let $a \in X$. There exist $y_1, y_2, \ldots$ in Y with $y_n \to a$. We claim that $y'_1, y'_2, \ldots$ is a Cauchy sequence. Indeed, let $\varepsilon > 0$. There is a $\delta > 0$ such that

$$x, y \in Y, \quad d(x, y) < \delta \quad \Longrightarrow \quad d'(x', y') < \varepsilon. \tag{$*$}$$

As $y_1, y_2, \ldots$ is Cauchy, there is an N such that for all $n, m \geq N$, we have $d(y_n, y_m) < \delta$, from which $d'(y'_n, y'_m) < \varepsilon$. Thus, $y'_1, y'_2, \ldots$ is Cauchy, hence convergent. We see that every $a \in X$ is associated with some $b \in X'$.

If $a \in X$ is associated with elements b and c of X', there exist y_1, $y_2, \ldots, z_1, z_2, \ldots \in Y$ such that $y_n \to a$, $z_n \to a$, $y'_n \to b$, and $z'_n \to c$. Then the sequence $y_1, z_1, y_2, z_2, \ldots$ converges to a. By what we have just seen, the sequence $y'_1, z'_1, y'_2, z'_2, \ldots$ must also converge. Hence, $b = c$. We see that every $a \in X$ is associated with at most one element of X'.

Define $g : X \to X'$ by the requirement that, for $a \in X$, $g(a)$ be *the* element of X' with which a is associated. It is clear that if f has a continuous extension $X \to X'$, this must be g. It is also clear that g is an extension of f. We are done if g is uniformly continuous.

Let $\varepsilon > 0$. Choose a positive δ such that $(*)$ holds. Let $x, y \in X$ and $d(x, y) < \delta$; we prove that $d'(g(x), g(y)) \leq \varepsilon$. (The uniform continuity follows.) There exist $x_1, x_2, \ldots, y_1, y_2, \ldots \in Y$ with $x_n \to x$, $y_n \to y$, $x'_n \to g(x)$, and $y'_n \to g(y)$. As $d(x_n, y_n) \to d(x, y)$ (see Exercise 5.G), we have $d(x_n, y_n) < \delta$ for all sufficiently large n, so that $d'(x'_n, y'_n) < \varepsilon$ for all sufficiently large n. But, $d'(x'_n, y'_n) \to d'(g(x), g(y))$, so $d'(g(x), g(y)) \leq \varepsilon$.

(ii) If f is Lipschitz with Lipschitz constant K, then in (∗) we can take
$\delta := K^{-1}\varepsilon$. It follows that g also has Lipschitz constant K. If f is
isometric, then [in the notation of the final part of the proof of (i)]
$d(x_n, y_n) = d'(x'_n, y'_n)$, so $d(x, y) = d'(g(x), g(y))$.   ∎

8.12
Let (X, d) be a metric space, preferably not complete. Let us call a metric
space (X', d') an "extension" of (X, d) if $X \subset X'$ and $d(x, y) = d'(x, y)$ for
all $x, y \in X$.

We wish to extend (X, d) to a complete metric space $(X^\sim, d^\sim)$. We also
wish $X^\sim$ to be no larger than is necessary. It would be nice if $X^\sim$ could be
made to consist solely of limits of Cauchy sequences in X, i.e., if X were
a dense subspace of $X^\sim$.

It turns out that such an $(X^\sim, d^\sim)$ can be constructed, but by marginally
moderating our wishes, we can save an amount of dull work. Instead of
wanting $(X^\sim, d^\sim)$ to be an extension of (X, d) itself, let us be satisfied if it
is an extension of a space that is isomorphic to (X, d). (See 5.4(vii).) As a
reward, we will obtain a uniqueness result: All spaces $(X^\sim, d^\sim)$ that meet
our wishes are mutually isomorphic.

8.13
We define formally: A *completion* of a metric space (X, d) is a pair consisting of a complete metric space $(X^\sim, d^\sim)$ and an isometry $j : X \to X^\sim$
such that $j(X)$ is dense in $X^\sim$. [Then, $X^\sim$ is an "extension" of the space $j(X)$
that is isomorphic to X.]

Theorem 8.14
Every metric space has a completion.

Proof
Let (X, d) be a metric space, $X \neq \varnothing$. With every $x \in X$ we connect a
function f_x on X by

$$f_x(t) := d(x, t) \qquad (t \in X).$$

[See Exercise 5.9(iv).] By the Second Triangle Inequality (5.4(i)), for all
$x, y \in X$ we have

$$|f_x(t) - f_y(t)| \leq d(x, y) \qquad (t \in X)$$

so that $f_x - f_y \in l^\infty(X)$ and $\|f_x - f_y\|_\infty \leq d(x, y)$. Better than that, we also
have $\|f_x - f_y\|_\infty \geq |f_x(y) - f_y(y)| = d(x, y)$. Thus,

$$\|f_x - f_y\|_\infty = d(x, y) \qquad (x, y \in X).$$

Now choose $e \in X$ and define $j : X \to l^\infty(X)$ by

$$j(x) := f_x - f_e \qquad (x \in X).$$

It is easy to see that j is an isometry: For all x and y,

$$d_\infty(j(x), j(y)) = \|(f_x - f_e) - (f_y - f_e)\|_\infty = \|f_x - f_y\|_\infty = d(x, y).$$

Let X' be the closure of $j(x)$ in $l^\infty(X)$, and d' the metric on X' induced by d_∞. By Theorem 7.7 and the completeness of $l^\infty(X)$, (X', d') is complete. ∎

As for the uniqueness:

Theorem 8.15
Let (X,d) be a metric space and let $((X_1,d_1),j_1)$ and $((X_2,d_2),j_2)$ be completions of (X,d). Then, (X_1,d_1) and (X_2,d_2) are isomorphic as metric spaces. More precisely, there exists an isometry j of X_1 onto X_2 such that $j \circ j_1 = j_2$.

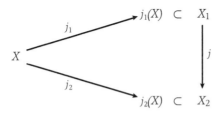

Proof
j_1 has an inverse $j_1^{-1} : j_1(X) \to X$. Then $j_2 \circ j_1^{-1}$ is an isometry $j_1(X) \to X_2$. By Theorem 8.11, $j_2 \circ j_1^{-1}$ extends to an isometry $j : X_1 \to X_2$. Trivially, $j \circ j_1 = j_2$. As X_1 is isomorphic to $j(X_1)$ and X_1 is complete, so is $j(X_1)$. Then $j(X_1)$ must be closed in X_2 (Exercise 7.E). But, $j(X_1) \supset j_2(X)$ and $j_2(X)$ is dense in X_2. Hence, $j(X_1) = X_2$, i.e., j is surjective. ∎

Extra: Spaces of Continuous Functions

For a nonempty metric space X, let us denote by $C(X)$ the set of all continuous functions $X \to \mathbb{R}$. This $C(X)$ is not just a set; it comes with a lot of structure. It is profitless to try and define the term "structure," but from some examples you will get the idea.

There is an addition: The sum of two continuous functions is a continuous function. More formally, if $f, g \in C(X)$, then $f + g \in C(X)$, where $f + g$ is the function

$$x \mapsto f(x) + g(x) \qquad (x \in X).$$

Similarly, there is a multiplication. If, for $f, g \in C(X)$, we let $f \bullet g$ denote the function

$$x \mapsto f(x)g(x) \qquad (x \in X),$$

then $f \bullet g \in C(X)$. (This $f \bullet g$ is called fg in the main text. We momentarily change the notation because further on it will be convenient to have a symbol for the multiplication at our disposal.)

Addition and multiplication are examples of structure on $C(X)$. The structure of a different character is the "ordering." For $f, g \in C(X)$, we write

$$f \le g \tag{1}$$

as an abbreviation for

$$f(x) \le g(x) \quad \text{for all} \quad x \in X.$$

$\le$ is a binary relation on $C(X)$: For certain combinations of f and g the formula (1) will be valid; for others, not.

Let us focus on $\bullet$ for a while. Consider the following question.

$$\text{How many elements } f \text{ exist} \tag{2}$$
$$\text{in } C(X) \text{ for which } f \bullet f = f?$$

There are at least two, namely the constant functions with values 0 and 1; so much we can say in general. Looking further, however, we see that the answer depends on X. It is "two" if, say, X is $\mathbb{R}$, but "four" if X is $[0, 1] \cup [2, 3]$. Thus, the answer tells us something about X.

Problem (as yet very vague): *Supposing we know everything about* $\bullet$ *that there is to be known, how much information about X can we deduce?*

Suppose we have two metric spaces, X and Y, and suppose there is given a "homeomorphism" $\varphi : X \to Y$; i.e., a bijective map $\varphi : X \to Y$ such that both φ and its inverse are continuous. To every function $f : X \to \mathbb{R}$ corresponds a function $\tilde{f} : Y \to \mathbb{R}$:

$$\tilde{f} := f \circ \varphi^{-1}.$$

Then, f is continuous if and only if $\tilde{f}$ is, and the correspondence

$$f \mapsto \tilde{f} \quad (f \in C(X))$$

is a bijection of $C(X)$ onto $C(Y)$. This bijection behaves very nicely relative to the multiplication. Indeed, for $f, g, h \in C(X)$ we have

$$f \bullet g = h \iff \tilde{f} \bullet \tilde{g} = \tilde{h}.$$

Returning to our question (2), we see that there are precisely as many functions in $C(X)$ that are equal to their own squares as there are in $C(Y)$. ($f \bullet f = f$ if and only if $\tilde{f} \bullet \tilde{f} = \tilde{f}$.)

More generally, every property of the multiplication in $C(X)$ is shared by the one in $C(Y)$.

This brings us to the concept of "isomorphism." For two metric spaces, X and Y, we say that $C(X)$ and $C(Y)$ are $\bullet$-*isomorphic* if there exists a bijection $f \mapsto \widehat{f}$ of $C(X)$ onto $C(Y)$ such that

$$f \bullet g = h \iff \widehat{f} \bullet \widehat{g} = \widehat{h}$$

for all f, g, h in $C(X)$. Then the multiplications of $C(X)$ and $C(Y)$ have all properties in common. As far as multiplication is concerned, you might call $C(X)$ and $C(Y)$ "congruent."

The problem we put in italics, above, can now be made slightly more precise: If $C(X)$ and $C(Y)$ are •-isomorphic, how closely must X and Y resemble each other? We have just seen that X need not be equal to Y: For $C(X)$ and $C(Y)$ to be •-isomorphic, it suffices that X and Y be homeomorphic. Remarkably, the converse is also true: *$C(X)$ and $C(Y)$ are •-isomorphic if and only if X and Y are homeomorphic.* In this sense, the multiplication on $C(X)$ contains much information about X.

The addition does not: One can show that $C(X)$ and $C(Y)$ are +-isomorphic if, for example, X is $[0, 1]$ and $Y = [0, 1] \cup [2, 3]$. For the ordering, the result is again positive. If we call $C(X)$ and $C(Y)$ $\leq$-*isomorphic* if there is a bijection $f \mapsto \hat{f}$ of $C(X)$ onto $C(Y)$ with

$$f \leq g \iff \hat{f} \leq \hat{g} \qquad (f, g \in C(X)),$$

then *$C(X)$ and $C(Y)$ are $\leq$-isomorphic if and only if X and Y are homeomorphic.*

Further Reading

Gillman, L. and M. Jerison, *Rings of Continuous Functions*, Springer-Verlag, Heidelberg, 1976.

Exercises

8.A. Which of the following sequences $(f_n)_{n \in \mathbb{N}}$ converge(s) uniformly? Make sketches.

(a) $\quad f_n(x) = \dfrac{x}{1 + nx}$

(b) $\quad f_n(x) = nx e^{-nx^2}$

(c) $\quad f_n(x) = n^{\frac{1}{2}} x(1-x^2)^n \qquad (0 \leq x \leq 1; \ n \in \mathbb{N}).$

(d) $\quad f_n(x) = nx(1-x^2)^{n^2}$

(e) $\quad f_n(x) = \dfrac{x^n}{1 + x^n}$

8.B. Define $f_n : \mathbb{R} \to \mathbb{R}$ by

$$f_n(x) = e^{-|x+n|} \qquad (x \in \mathbb{R}).$$

Show that the sequence $(f_n)_{n \in \mathbb{N}}$ converges pointwise but *not* uniformly, and that the limit function *is* continuous.

8.C. Let $f, f_1, f_2, \ldots$ and $g, g_1, g_2, \ldots$ be functions on a set S; let $f_n \to f$ uniformly
and $g_n \to g$ uniformly.
   (i) Show that for all $\lambda, \mu \in \mathbb{R}$,

$$\lambda f_n + \mu g_n \to \lambda f + \mu g \quad \text{uniformly.}$$

   (ii) Suppose there exists a number K such that $|f_n(s)| \leq K$ and $|g_n(s)| \leq K$
for all n and s. Prove that

$$f_n g_n \to fg \quad \text{uniformly.}$$

   [The boundedness condition mentioned in (ii) is not redundant: Taking
$S = \mathbb{R}$ and $f_n(s) = \frac{1}{n}$, $f(s) = 0$, and $g_n(s) = g(s) = s$, one has $f_n \to f$ and
$g_n \to g$ uniformly but not $f_n g_n \to fg$ uniformly.]

8.D. Let (X, d) be a metric space and suppose d is bounded, i.e., there exists a
number M such that

$$d(x, y) \leq M \quad \text{for all } x, y \in X.$$

   In Exercise 5.K, for every nonempty subset A of X we have made a function
$f_A : X \to \mathbb{R}$ by

$$f_A(x) := \inf\{d(x, a) : a \in A\};$$

and we have seen that f_A is continuous. It is clear that

$$f_A(x) \leq M \quad \text{for all } A \text{ and } x.$$

   Let $\mathcal{C}$ be the collection of all nonempty closed subsets of X.
   (i) Show (using Exercise 6.E?) that the formula

$$\delta(A, B) := \|f_A - f_B\|_\infty \qquad (A, B \in \mathcal{C})$$

   defines a metric δ on $\mathcal{C}$. (This δ is called the *Hausdorff metric*.)
   (ii) Show that the map

$$a \mapsto \{a\}$$

   is an isometry of X into $\mathcal{C}$.

8.E. Let (X, d) be a metric space; let $f : X \to X$. It is easy to get the impression
that f is already a contraction if

$$x, y \in X, \ x \neq y \ \implies \ d\big(f(x), f(y)\big) < d(x, y) \qquad (*)$$

   Take $X := [1, \infty)$, $f(x) = x + \frac{1}{x}$ $(x \in X)$.
   (i) Show that X is complete. (Euclidean metric.)
   (ii) Show that f maps X into X and has the property $(*)$.
   (iii) Show that f has no fixed point. [Then by (i) and the Contraction
Principle 7.14, f cannot be a contraction.]

8.F. By c_0 we indicate the set of all number sequences that converge to 0. We
know that all convergent sequences are bounded (Theorem 3.6), so c_0 is a
subset of $l^\infty(\mathbb{N})$.
   Show that c_0 is d_∞-closed in $l^\infty(\mathbb{N})$.
   Show that the metric space (c_0, d_∞) is complete.

8.G. By c we indicate the set of all convergent number sequences. Convergent sequences are bounded (Theorem 3.6), so c is a metric space under d_∞.

For $x = (x_1, x_2, \ldots) \in c$, put

$$L(x) := \lim_{n \to \infty} x_n;$$

then L is a function $c \to \mathbb{R}$. Prove that L is d_∞-continuous. [See 5.1(iv).]

8.H. For $f \in C[0, 1]$, let $M(f)$ be the largest value taken by f.

Show that

$$|M(f) - M(g)| \leq \|f - g\|_\infty \qquad (f, g \in C[0, 1])$$

and that the function $M : C[0, 1] \to \mathbb{R}$ is d_∞-continuous. [See 5.1(vi).]

8.I. We consider $C[0, 1]$, endowed with the metric d_∞.

For every interval $I \subset [0, 1]$, let $A[I]$ be the set of all continuous functions $f : [0, 1] \to \mathbb{R}$ that are increasing on I, i.e., for which

$$x, y \in I, \quad x \leq y \quad \Longrightarrow \quad f(x) \leq f(y),$$

and $B[I]$ the set of all continuous functions on $[0, 1]$ that are decreasing on I.

(i) Show that for every interval $I \subset [0, 1]$, the sets $A[I]$ and $B[I]$ are closed and have empty interior.

(ii) Let $I_1, I_2, \ldots$ be the intervals $[0, \frac{1}{2}], [\frac{1}{2}, 1], [0, \frac{1}{3}], [\frac{1}{3}, \frac{2}{3}], [\frac{2}{3}, 1], [0, \frac{1}{4}], [\frac{1}{4}, \frac{2}{4}], \ldots$. Deduce from the Baire Category Theorem that

$$C[0, 1] \neq \bigcup_n A[I_n] \cup \bigcup_n B[I_n].$$

(iii) Now show that *there exists a continuous function on $[0,1]$ that is neither increasing nor decreasing on any subinterval of $[0, 1]$.*

9

CHAPTER

Sequential Compactness

9.1

For subsets of $\mathbb{R}$ and $\mathbb{R}^2$, we have defined sequential compactness in Chapter 3. The same definition is meaningful in the setting of metric spaces in general.

Let (X, d) be a metric space. We call a subset Y of X *sequentially compact* if every sequence of elements of Y has a subsequence that converges to an element of Y. The last clause, "to an element of Y," is vital: In $\mathbb{R}$, the interval $(0, 1]$ is not sequentially compact, as the sequence $1, \frac{1}{2}, \frac{1}{3}, \ldots$ has no subsequence with a limit in $(0, 1]$.

In 6.17 we have seen that the property of openness is not "intrinsic": A set is (or is not) open as a subset of a metric space. $[0, 1]$ is open in $[0, 1]$ but not in $\mathbb{R}$. Sequential compactness is different, as is apparent from the definition. The statements "$[0, 1]$ is sequentially compact in $[0, 1]$" and "$[0, 1]$ is sequentially compact in $\mathbb{R}$" have the same meaning.

In Example 3.26 we have considered examples of sequentially compact subsets of $\mathbb{R}^2$. It follows from the Bolzano-Weierstrass Theorem that every closed bounded subset of $\mathbb{R}$ (such as the Cantor set; see Example 6.10) is sequentially compact. Other examples will follow in Examples 9.7(ii) and 9.11 and Exercises 9.F and 9.H. First, a few simple remarks, either very elementary or immediate generalizations of earlier results.

Theorem 9.2

Every closed subset of a sequentially compact metric space is sequentially compact.

Proof

Let X be a sequentially compact metric space and Y a closed subset of X. Every sequence in Y has a subsequence converging to a point of X. As Y is (sequentially) closed, this limit lies in Y. ∎

Theorem 9.3

On a nonempty sequentially compact metric space, every continuous function attains a largest value and is bounded.

Proof

Precisely as in Theorem 3.9, Corollary 3.10, and Theorem 3.24. ∎

Theorem 9.4

On a sequentially compact metric space, every continuous function is uniformly continuous.

After Theorems 3.13 and 3.25, the reader will be able to supply both a reasonable definition of "uniformly continuous" and a proof of the theorem.

Theorem 9.5

Every sequentially compact subset of a metric space X is closed in X.

Proof

Let X be a metric space and let $Y \subset X$ be sequentially compact. We show Y to be sequentially closed. (See 6.8.) Let $y_1, y_2, \ldots \in Y$, $a \in X$, and $y_n \to a$. The sequence $y_1, y_2, \ldots$ has a subsequence converging to a point of Y. But every subsequence of $y_1, y_2, \ldots$ converges to a [5.4(iii)]. Hence, $a \in Y$. ∎

Theorem 9.6

A subset of $\mathbb{R}$ or $\mathbb{R}^N$ is sequentially compact if and only if it is both closed and bounded.

Proof for $\mathbb{R}^N$

(I) Let $Y \subset \mathbb{R}^N$ be sequentially compact. We have just seen that Y must be closed. The function $x \mapsto \|x\|$ is (sequentially) continuous on $\mathbb{R}^N$; then, by Theorem 9.3, it is bounded on Y. It follows that Y is a bounded set.

(II) Conversely, suppose $Y \subset \mathbb{R}^N$ is closed and bounded. Take a sequence in Y. If we can show that it has a convergent subsequence, then, by the closedness of Y, the limit of such a subsequence must lie in Y, and we are done. Thus, it suffices to prove that in $\mathbb{R}^N$ every bounded sequence has a convergent subsequence.

That has been done in Theorem 3.18 for $N = 2$. It is easy to see how repeating the argument given there yields a proof for arbitrary N. (If you are not convinced, another proof will follow in Exercise 9.A.)

Examples 9.7

(i) Many closed bounded subsets of $\mathbb{R}$ and $\mathbb{R}^N$ are easily recognizable as such. The obvious examples are the closed bounded intervals of $\mathbb{R}$, and the unit disk and the unit square in $\mathbb{R}^2$; we have known of their sequential compactness since Section 3.

(ii) In $\mathbb{R}^3$, the "unit sphere" $S := \{x \in \mathbb{R}^3 : \|x\| = 1\}$ is bounded and (sequentially) closed, hence sequentially compact under the Euclidean metric d_E. In 5.3(iv), we have considered a different metric on S, namely

$$d(x, y) := \cos^{-1}\langle x, y \rangle.$$

We can now show that S is also sequentially compact relative to this metric d. Indeed, let $x_1, x_2, \ldots$ be a sequence in S. This sequence has a subsequence $y_1, y_2, \ldots$ that d_E-converges to some $y \in S$. Then, $\langle y_n, y \rangle \to \langle y, y \rangle = 1$, so $d(y_n, y) = \cos^{-1}\langle y_n, y \rangle \to \cos^{-1} 1 = 0$ and the sequence $y_1, y_2, \ldots$ also d-converges to y.

(iii) It is, of course, not to be expected that in *every* metric space, all closed bounded subsets are sequentially compact. In fact, if d_0 is the trivial metric on some set X [see 5.3(vi)], then every subset of X is closed [Example 6.9(iii)] and bounded, but only the finite subsets are sequentially compact.

9.8

The next few pages deal with collections of sets. The reader may wish to look back at 6.11 and 6.12 for the relevant terminology and notations such as $\in$ and $\sqsubset$.

Let X be a metric space—or, for that matter, any set. A collection ω of subsets of X is said to *cover* X, or to be a *cover* of X, if every point of X lies in a set that belongs to ω, i.e., if $X = \bigcup \omega$ (e.g., the intervals of length 1 cover $\mathbb{R}$.)

A cover of X is *finite*, simply if only finitely many sets belong to it.

If ω and ω' are covers of X, we call ω' a *subcover* of ω if $\omega' \sqsubset \omega$ (e.g., let X be $\mathbb{R}$, let ω be the collection of all intervals, and let ω' be the collection of all closed intervals of length 1.)

9.9

A metric space X is called *totally bounded* (or *precompact*) if for every $\varepsilon > 0$, X can be covered by finitely many subsets that each have diameter $\leq \varepsilon$. (See 7.2.)

Observe: If X is totally bounded, then so is every subset of X.

A totally bounded space X is bounded: Cover X by nonempty sets $X_1, \ldots, X_N$ with diam $X_n \leq 1$ for each n; choose $x_1 \in X_1, \ldots, x_N \in X_N$. Then, for all $x, y \in X$,

$$d(x, y) \leq 2 + \max\{d(x_n, x_m) : n, m = 1, \ldots, N\}.$$

In $\mathbb{R}^N$, all bounded sets are totally bounded, as is easy to see. This is not the case in, say, $l^\infty(\mathbb{N})$ with the metric d_∞: Let X be $\{f \in l^\infty(\mathbb{N}) : \|f\|_\infty \leq 1\}$. Then, X is bounded, but it contains the indicator functions of all one-point subsets of $\mathbb{N}$, and no two of these indicator functions can lie in one set with diameter less than 1.

Theorem 9.10
A metric space (X,d) is sequentially compact if and only if it is complete and totally bounded.

Proof

(I) Assume (X, d) is sequentially compact.

Let $x_1, x_2, \ldots$ be a Cauchy sequence:

$$d(x_n, x_m) < \varepsilon_n \quad \text{if} \quad m \geq n,$$

where $\varepsilon_n \to 0$. By compactness, there is a strictly increasing $\alpha : \mathbb{N} \to \mathbb{N}$ such that $x_{\alpha(n)} \to a$ for some $a \in X$. As $\alpha(n) \geq n$ for all n, we have $d(x_n, x_{\alpha(n)}) < \varepsilon_n$ and thereby

$$d(x_n, a) \leq \varepsilon_n + d(x_{\alpha(n)}, a).$$

Hence, $x_n \to a$. This proves the completeness.

Let $\varepsilon > 0$. *Suppose* X cannot be covered by finitely many sets of diameter ε. Then X cannot be covered by finitely many balls of radius $\varepsilon/2$. It follows that we can make an infinite sequence $a_1, a_2, \ldots$ in X such that

$$a_2 \notin B_{\varepsilon/2}(a_1),$$

$$a_3 \notin B_{\varepsilon/2}(a_1) \cup B_{\varepsilon/2}(a_2),$$

etc.

For all m and n with $m \neq n$, we have $d(a_m, a_n) > \varepsilon/2$. Then, the sequence $a_1, a_2, \ldots$ cannot have a convergent subsequence. *Contradiction.*

(II) Assume that (X, d) is complete and totally bounded. Let $x_1, x_2, \ldots \in X$. We wish to obtain a convergent subsequence. We may assume (?) that the set $S := \{x_1, x_2, \ldots\}$ is infinite.

Cover X by finitely many subsets with diameter at most equal to 1. At least one of these must contain infinitely many elements of S. Hence, S has an infinite subset S_1 for which diam·$S_1 \leq 1$. Similarly,

S_1 has an infinite subset S_2 with diam $S_2 \leq \frac{1}{2}$, and so forth. Thus, there is a sequence of infinite sets

$$S \supset S_1 \supset S_2 \supset \cdots$$

with diam $S_p \leq p^{-1}$ for each p. Choose $\alpha(1) \in \mathbb{N}$ with $x_{\alpha(1)} \in S_1$. As S_2 is an infinite set, there are infinitely many indices n with $x_n \in S_2$; choose $\alpha(2) \in \mathbb{N}$ such that $x_{\alpha(2)} \in S_2$ and $\alpha(2) > \alpha(1)$. Similarly, take an $\alpha(3)$ with $x_{\alpha(3)} \in S_3$ and $\alpha(3) > \alpha(2)$; and so on. Then, $x_{\alpha(1)}, x_{\alpha(2)}, \ldots$ is a subsequence of $x_1, x_2, \ldots$. It is Cauchy since $d(x_{\alpha(n)}, x_{\alpha(m)}) \leq p^{-1}$ as soon as $n, m \geq p$. By completeness, it converges. ∎

Example 9.11

Let $l^\infty(\mathbb{N})$ be the set of all bounded functions on $\mathbb{N}$. In 8.4, we imposed on $l^\infty(\mathbb{N})$ the metric d_∞ that determines uniform convergence. Consider the subset

$$K := \{f \in l^\infty(\mathbb{N}) : 0 \leq f(n) \leq n^{-1} \text{ for every } n \in \mathbb{N}\}.$$

We claim that under the metric induced by d_∞, this K is complete and totally bounded, hence sequentially compact. Indeed, $l^\infty(\mathbb{N})$ is complete (Theorem 8.7) and K clearly is (sequentially) closed in $l^\infty(\mathbb{N})$; so, by Theorem 7.7, K is complete. As for the total boundedness, let $N \in \mathbb{N}$; we show that finitely many balls of radius $1/N$ cover K. Consider the set K_N of all $g \in K$ with

$$g(n) \in \left\{0, \frac{1}{N}, \frac{2}{N}, \ldots, \frac{N-1}{N}, 1\right\} \quad \text{for every } n.$$

If $g \in K_N$, then $g(n) = 0$ as soon as $n > N$, so K_N is a finite set. The balls $B_{1/N}(g)$ with $g \in K_N$ cover K.

Corollary 9.12

Every sequence of functions $\mathbb{N} \to [0, 1]$ has a subsequence that converges pointwise.

Proof

Let $f_1, f_2, \ldots$ be functions of $\mathbb{N}$ into $[0, 1]$. Define functions $f_1^*, f_2^*, \ldots$ on $\mathbb{N}$ by

$$f_i^*(n) := n^{-1} f_i(n) \quad (n \in \mathbb{N}).$$

By Example 9.11, the sequence $f_1^*, f_2^*, \ldots$ has a subsequence $f_{\alpha(1)}^*, f_{\alpha(2)}^*, \ldots$ that converges in the sense of d_∞, i.e., uniformly, and therefore pointwise. Then $f_{\alpha(1)}, f_{\alpha(2)}, \ldots$ converges pointwise. ∎

9.13

For another description of sequential compactness we need some new terminology. Let X be a set and ω a nonempty collection of subsets of X.

We say that ω is *finitely bound* (or, has the *finite intersection property*) if

$$N \in \mathbb{N}, \ A_1, \ldots, A_N \in \omega \implies A_1 \cap \ldots \cap A_N \neq \varnothing.$$

Obviously, this is the case if ω has nonempty intersection; that is, if there is a point of X that lies in every set belonging to ω. The converse is false: If $X = \mathbb{R}$ and ω consists of the intervals $[x, \infty)$ $(x \in \mathbb{R})$, then ω is finitely bound but has empty intersection.

Observe: If ω is finitely bound, then $\varnothing$ cannot belong to ω.

Theorem 9.14

Let X be a metric space. Then, conditions (α), (β), and (γ) are equivalent.

(α) *X is sequentially compact.*
(β) *Every finitely bound collection of closed subsets of X has nonempty intersection.*
(γ) *Every cover of X by open sets has a finite subcover. (See 9.8.)*

Proof

Let d be the metric of X.

$(\alpha) \implies (\beta)$ Let φ be a finitely bound collection of closed subsets of X.

Take $\varepsilon > 0$. As X is totally bounded (Theorem 9.10), we can cover X by finitely many balls of radius ε, say, $X_1, \ldots, X_M$. We claim that there must be an $m \in \{1, \ldots, M\}$ with

$$X_m \text{ intersects every set of the collection } \varphi.$$

Otherwise, for every m, there is an $A_m \in \varphi$ such that $X_m \cap A_m = \varnothing$. Then $A_1 \cap \cdots \cap A_M$ contains a point z. As z lies in every A_m, it cannot lie in any X_m. *Contradiction.*

Thus, for each n in $\mathbb{N}$, there is an x_n in X with

$$B_{1/n}(x_n) \text{ intersects every set of the collection } \varphi. \tag{$*$}$$

We obtain a sequence $x_1, x_2, \ldots$. This sequence has a subsequence $x_{\alpha(1)}, x_{\alpha(2)}, \ldots$ converging to some $a \in X$. We claim that a lies in every set that belongs to φ.

Indeed, let $A \in \varphi$. We show that a is adherent to A (so that $a \in A$ since A is closed). Let $\varepsilon > 0$. Choose $n \in \mathbb{N}$ so large that $x_{\alpha(n)} \in B_{\varepsilon/2}(a)$ and $\alpha(n)^{-1} < \varepsilon/2$. Then,

$$B_{1/\alpha(n)}(x_{\alpha(n)}) \subset B_\varepsilon(a),$$

so that $B_\varepsilon(a) \cap A \neq \varnothing$ by $(*)$.

$(\beta) \implies (\alpha)$ Take a sequence $x_1, x_2, \ldots$ in X. For $n \in \mathbb{N}$, let A_n be the closure of $\{x_n, x_{n+1}, \ldots\}$. Then, $\{A_1, A_2, \ldots\}$ is a collection of closed subsets of X. It is finitely bound since $A_1 \supset A_2 \supset \cdots$ and no A_n is empty. Hence, there is a point a of X that lies in every A_n. We proceed to make a subsequence of $x_1, x_2, \ldots$ that converges to a.

As $a \in A_1$, there is an $\alpha(1) \in \mathbb{N}$ with $d(x_{\alpha(1)}, a) < 1$. As $a \in A_{\alpha(1)+1}$, there is an $\alpha(2) \in \mathbb{N}$ with $\alpha(2) \geq \alpha(1) + 1$ and $d(x_{\alpha(2)}, a) < \frac{1}{2}$. Similarly, there is an $\alpha(3) \in \mathbb{N}$ with $\alpha(3) \geq \alpha(2) + 1$ and $d(x_{\alpha(3)}, a) < \frac{1}{3}$, and so forth. We obtain a subsequence $x_{\alpha(1)}, x_{\alpha(2)}, \ldots$ of $x_1, x_2, \ldots$ with $x_{\alpha(n)} \to a$.

$(\beta) \implies (\gamma)$ Let ω be a cover of X by open sets; we look for a finite subcover. The sets

$$X \backslash U \qquad (U \sqsubseteq \omega)$$

form a collection φ of closed sets. For every $x \in X$, there is a $U \sqsubseteq \omega$ with $x \in U$, and, therefore, there is an $A \sqsubseteq \varphi$ with $x \notin A$. Thus, φ has empty intersection. By condition (β), φ cannot be finitely bound. Hence, there exist $U_1, \ldots, U_N \sqsubseteq \omega$ with

$$(X \backslash U_1) \cap \cdots \cap (X \backslash U_N) = \varnothing,$$

i.e., $X = U_1 \cup \ldots \cup U_N$ and we have our finite subcover.

The proof of the implication $(\gamma) \implies (\beta)$ we leave confidently to the reader.   ∎

Extra: The p-adic Numbers

By $\mathbb{Z}_{10}$ we mean the set of all sequences $x_1, x_2, x_3, \ldots$ with $x_n \in \{0, 1, 2, 3, \ldots, 8, 9\}$ for every n, such as the sequences of the decimals of e and $1/7$:

$$7, 1, 8, 2, 8, \ldots \qquad \text{and} \qquad 1, 4, 2, 8, 5, \ldots.$$

The elements of $\mathbb{Z}_{10}$ are called 10-*adic integers*. With special purposes in mind, we change the notation: Instead of

$$x_1, x_2, x_3, \ldots,$$

we write

$$\ldots x_3 x_2 x_1.$$

Thus, $a := \ldots 82817$ and $b := \ldots 58241$ are 10-adic integers.

We introduce an addition and a multiplication in $\mathbb{Z}_{10}$ like this:

```
      .   .   . 8  2  8  1  7              .   .   . 8  2  8  1  7
  +   .   .   . 5  8  2  4  1          x   .   .   . 5  8  2  4  1
  ─────────────────────────          ─────────────────────────────
      .   .   . 4  1  0  5  8              .   .   . 8  2  8  1  7/
                                          .   .   . 3  1  2  6  8/7
                                          .   .  6  5  6  3  4/9
                                          .  6  2  5  3  6/8
                                          .   .   .   .   . 5/4
                                          .   .   .   .   .   /
```

You get the idea. Just to show that we are not cheating, we describe a more exact way to give the definitions.

For $x \in \mathbb{Z}_{10}$ and $n \in \mathbb{N}$, by $[x]_n$ we indicate the positive integer that is given by the final n digits of x. Thus, with a and b as above,

$$[a]_3 = 817, \quad [b]_3 = 241.$$

Now compare the above multiplication of a and b with the ordinary multiplication of the integers $[a]_3$ and $[b]_3$:

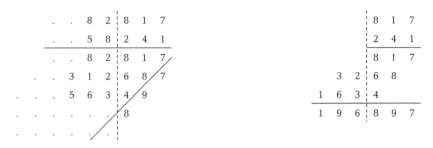

You see what happens: If c is the (somewhat informally defined) product of a and b, then

$$[c]_3 = [a]_3[b]_3 + (\text{a multiple of } 10^3).$$

This observation leads to a formal definition of the multiplication: The product of two 10-adic integers x and y is the 10-adic integer z satisfying, for all $N \in \mathbb{N}$,

$$[z]_n = [x]_n[y]_n + (\text{a multiple of } 10^n).$$

(It is easy to see that there is precisely one such z.) Similarly, the sum of x and y is the 10-adic integer u with

$$[u]_n = [x]_n + [y]_n + (\text{a multiple of } 10^n)$$

for all n.

You can now prove that these operations satisfy formulas completely analogous to (a), (b), (e), (f), and (g) of our Axiom I in 2.2. (Commutativity, associativity, and distributivity.)

With an ordinary non-negative integer p we associate a 10-adic integer $\overline{p}$ as follows: Write p in the usual way as a finite string of digits, then let it be preceded by an infinite string of zeros. Thus,

$$\overline{385} = \ldots 000000385.$$

Then, for all $x \in \mathbb{Z}_{10}$

$$x + \overline{0} = x, \qquad x\overline{1} = x,$$

analogous to (c) and (h) of Axiom I. You will readily see that we have the natural analog of (d): For every x, there is a unique $-x$ with $x + (-x) = \overline{0}$, e.g.,

$$- \ldots 82817 = \ldots 17183.$$

For every p in $\{0, 1, 2, \ldots\}$, we have made a 10-adic integer $\bar{p}$. We can extend the definition to all integers by

$$\bar{p} := -\overline{-p} \qquad (p = -1, -2, \ldots).$$

Then

$$\overline{p+q} = \bar{p} + \bar{q}, \quad \overline{pq} = \bar{p}\,\bar{q} \qquad (p, q \in \mathbb{Z}).$$

In $\mathbb{Z}_{10}$, we have no analog for (i) of Axiom I, as there is no 10-adic integer y with $\overline{10}\, y = \bar{1}$.

You may be wondering what all this has to do with metric spaces. The connection is the following. We define a metric d on $\mathbb{Z}_{10}$ by setting, for $x, y \in \mathbb{Z}_{10}$,

$$d(x, y) := 0 \qquad \text{if} \quad x = y,$$

$$d(x, y) := m^{-1} \quad \text{if} \quad x-y \quad \text{ends in precisely } m-1 \text{ zeros.}$$

Then for $p, q \in \mathbb{Z}$, $d(\bar{p}, \bar{q})$ is precisely the "10-adic distance" between p and q in the sense of Example 5.5(v); in other words, the map $p \mapsto \bar{p}$ is an isometry relative to the 10-adic metric on $\mathbb{Z}$. As a metric space, $\mathbb{Z}_{10}$ is complete and even sequentially compact. Indeed, it is a completion of $\mathbb{Z}$ (under the 10-adic metric).

It is not difficult to see that d-convergence is "coordinatewise": If $x, x_1, x_2, \ldots \in \mathbb{Z}_{10}$, then

$$d(x_n, x) \to 0 \iff [x_n]_1 = [x]_1 \text{ for large } n, \text{ and}$$

$$[x_n]_2 = [x]_2 \text{ for large } n, \text{ and}$$

$$[x_n]_3 = [x]_3 \text{ for large } n,$$

etc.

It follows that addition and multiplication are continuous. Note that $d(x, y) = d(x+z, y+z)$ for all x, y, and z. In particular,

$$d(x, y) = d(x-y, \bar{0}) \qquad (x, y \in \mathbb{Z}_{10}).$$

We have seen that division does not work well in $\mathbb{Z}_{10}$: You cannot divide $\bar{1}$ by $\overline{10}$ (but you can divide by $\bar{7}$). Something can be done by extending $\mathbb{Z}_{10}$ to $\mathbb{Q}_{10}$, the system of "10-adic numbers." A 10-*adic number* is obtained by taking a 10-adic integer and adding a "decimal point" and finitely many digits, e.g.,

$$\ldots 82817.23 \qquad \text{or} \qquad \ldots 28514.4 \,.$$

It is then understood that $\ldots 514.3$ and $\ldots 514.300$ are the same 10-adic number, and also that $\ldots 2817 = \ldots 2817.0$ (so that $\mathbb{Z}_{10} \subset \mathbb{Q}_{10}$).

Addition and multiplication have natural extensions to $\mathbb{Q}_{10}$, and the 10-adic number $\ldots 0000.1$ is the inverse of $\overline{10}$. The metric can also be extended; for instance, by defining for $x, y \in \mathbb{Q}_{10}$,

$$d(x, y) := d(x - y, 0) \quad \text{if} \quad x - y \in \mathbb{Z}_{10},$$

$$d(x, y) := 1 \qquad\qquad \text{if} \quad x - y \notin \mathbb{Z}_{10}.$$

Then, as in $\mathbb{Z}_{10}$, addition and multiplication are continuous.

In $\mathbb{Q}_{10}$, one can divide by $\overline{10}$ and even, less obviously, by $\overline{p}$ for every $p \in \mathbb{Z}$, $p \neq 0$, but not by every nonzero 10-adic number. However, we can get farther by discarding the decimal system. We know how to calculate in, say, base 7. (What makes 7 more interesting than 10 is that it is a prime number.) Let $\mathbb{Z}_7$ be the set of all strings $\ldots x_3 x_2 x_1$ of elements of $\{0, 1, 2, 3, 4, 5, 6\}$. Introduce addition and multiplication in $\mathbb{Z}_7$ as we did in $\mathbb{Z}_{10}$, but now working in base 7. Next, make the set $\mathbb{Q}_7$ of all "7-adic numbers." This $\mathbb{Q}_7$ can be turned into a metric space with continuous addition and multiplication having all the desirable properties *and* admitting division by every nonzero 7-adic number.

This construction opens an entire new world. In $\mathbb{Q}_7$, you can study convergence of series and differentiation of functions, and you can solve differential equations; in short, you can build up a new calculus. In many respects, this calculus is just like the one you know (the Chain Rule holds, for instance), in others it is weird. (Some nonconstant functions have derivative zero everywhere, and $\sum n!$ converges.)

In the same way, for every prime number p, you can make the system $\mathbb{Q}_p$ of p-adic numbers and do p-adic analysis.

Further Reading

Bachman, G., *Introduction to p-adic Numbers and Valuation Theory*, Academic Press, New York, 1964.

Exercises

9.A. Let Y be a closed bounded subset of $\mathbb{R}^N$. Show that Y is complete and totally bounded, hence sequentially compact. (This is the proof we announced in Theorem 9.6.)

9.B. Let (X', d') and $(X," d'')$ be metric spaces and $X := X' \times X''$. In Exercise 5.H, we have defined a metric d on X by

$$d(x, y) := d'(x', y') + d''(x," y'')$$

for $x = (x', x'') \in X$ and $y = (y', y'') \in X$. Prove: If (X', d') and $(X,'' d'')$ are sequentially compact, then so is (X, d). (Hint: look at the proof of Theorem 3.18.)

9.C. Let (X, d) be sequentially compact. Show that there exist $a, b \in X$ such that

$$d(a, b) \geq d(x, y) \quad \text{for all} \quad x, y \in X.$$

9.D. Let A be a subset of a *complete* metric space X. Prove that

$$A \text{ is precompact} \iff A^- \text{ is compact.}$$

9.E. Let (X, d) be a totally bounded metric space.
   (i) Prove that there exists a sequence $B_1, B_2, \ldots$ of balls with the following property:

For every open set U and every $a \in U$,

there is an n such that $a \in B_n \subset U$.

Hint: For every $N \in \mathbb{N}$, choose a finite set X_N such that the balls $B_{1/N}(x)$ with $x \in X_N$ cover X. Show that for all $a \in X$ and $r > 0$, there exist an $N \in \mathbb{N}$ and an $x \in X_N$ for which $a \in B_{1/N}(x) \subset B_r(a)$.
   (ii) Take $B_1, B_2, \ldots$ as in (i). For each n, let a_n be a center of B_n and define a function f_n on X by

$$f_n(x) := d(x, a_n) \qquad (x \in X).$$

[By Example 5.9(iv), f_n is continuous.] Prove that, for $x, y \in X$,

$$x = y \iff f_n(x) = f_n(y) \text{ for every } n.$$

9.F. Prove the following theorem. *If X is a sequentially compact metric space and f is a continuous map of X into a metric space Y, then $f(X)$ is sequentially compact and a closed subset of Y.*

9.G. Take K as in Example 9.11. We know that uniform convergence implies pointwise convergence. Show that in K, the converse is also true. [Hint: If $f, g \in K, N \in \mathbb{N}$, and

$$|f(1)-g(1)| \leq N^{-1}, |f(2)-g(2)| \leq N^{-1}, \ldots, |f(N)-g(N)| \leq N^{-1},$$

then $d_\infty(f, g) \leq N^{-1}$.)

9.H. Consider the set $\mathcal{P}(\mathbb{N})$ of all subsets of $\mathbb{N}$ under the metric d introduced in Example 5.3(v). For $A \in \mathcal{P}(\mathbb{N})$, we define the function $f_A : \mathbb{N} \to \mathbb{R}$ by

$$f_A(n) := n^{-1}\mathbf{1}_A(n) \qquad (n \in \mathbb{N}).$$

Show that $A \mapsto f_A$ is an isometry of $\mathcal{P}(\mathbb{N})$ into the set K mentioned in Example 9.11. Prove that $\mathcal{P}(\mathbb{N})$ *is sequentially compact under the metric d.*

9.I. (On $\mathcal{P}(\mathbb{N})$ and the Cantor set.)
   In Example 5.3(v), we defined a metric d on $\mathcal{P}(\mathbb{N})$, the collection of all subsets of $\mathbb{N}$ (see the previous exercise also) and in Example 6.10, we considered the Cantor set C. We will now see that $\mathcal{P}(\mathbb{N})$ *and C are homeomorphic.*

We introduce a collection of intervals:

$I := [0, 1]$;

if $J = [a, b]$, then $J_0 := \left[a, a + \dfrac{b-a}{3}\right]$, $J_1 := \left[b - \dfrac{b-a}{3}, b\right]$.

Thus, the set C_1 of Example 6.10 is $I_0 \cup I_1$, C_2 is $I_{00} \cup I_{01} \cup I_{10} \cup I_{11}$, and so forth. If $i_1, i_2, \ldots$ is any sequence of zeros and ones, then the intervals

$$I_{i_1}, I_{i_1 i_2}, I_{i_1 i_2 i_3}, \ldots$$

form a shrinking sequence whose intersection consists of one element of the Cantor set.

ii)

(i) Show that for every subset A of $\mathbb{N}$, there exists a unique real number x_A such that for every $m \in \mathbb{N}$,

$$i_1 = \mathbf{1}_A(1),\, i_2 = \mathbf{1}_A(2), \ldots, i_m = \mathbf{1}_A(m) \quad \Longrightarrow \quad x_A \in I_{i_1 i_2 \ldots i_m}.$$

Show that the formula $A \mapsto x_A$ describes a bijection $\mathcal{P}(\mathbb{N}) \to C$.
Prove: If $A, B \in \mathcal{P}(\mathbb{N})$ and $m \in \mathbb{N}$, then

$$d(A, B) < m^{-1} \quad \Longleftrightarrow \quad \text{there exist } i_1, i_2, \ldots, i_m$$

$$\text{with } x_A, x_B \in I_{i_1 i_2 \ldots i_m},$$

and

$$|x_A - x_B| < 3^{-m} \quad \Longrightarrow \quad d(A, B) < m^{-1} \quad \Longrightarrow \quad |x_A - x_B| \leq 3^{-m}.$$

Deduce that both the map $A \mapsto x_A$ and its inverse are continuous. Show that $\left(\mathcal{P}(\mathbb{N}), d\right)$ is sequentially compact.

(ii) Show that for all $i_1, i_2, \ldots, i_m$ the left end point of $I_{i_1 i_2 \ldots i_m}$ is

$$2(i_1 3^{-1} + i_2 3^{-2} + \cdots + i_m 3^{-m}).$$

Now prove

$$x_A = 2 \sum_{n=1}^{\infty} \mathbf{1}_A(n) 3^{-n} \quad \left(A \in \mathcal{P}(\mathbb{N})\right).$$

Is $\frac{1}{4}$ an element of the Cantor set?

10

Convergent Nets

CHAPTER

10.1

So far, we have been studying aspects of metric spaces such as convergence, continuity, open and closed sets, completeness, and sequential compactness. We know that sometimes two distinct metrics on a certain set determine the same notion of convergence. In 6.20, we have called such metrics "similar." A good example is formed by the Euclidean and the $\tan^{-1}$ metrics on $\mathbb{R}$; see Example 5.5(ii).

In Theorem 6.21 we have seen that two similar metrics lead to the same concepts of openness and closedness. Notions such as convergence, openness, and closedness are called "topological." Sequential compactness trivially belongs to the same category. Completeness does not: $\mathbb{R}$ is complete under the Euclidean metric but not under the $\tan^{-1}$ metric (Exercise 7.A).

When one studies topological properties of metric spaces, the metric is, to some extent, extraneous. Like a coordinate system in a vector space, it is occasionally helpful for carrying out a proof but often not really relevant to the matter at hand. It turns out that "topology" without explicit mention of the metric is surprisingly simple, once you get used to the abstraction.

10.2

How does one begin with "topology without a metric"? A metric on X basically is a list giving you for every pair of points of X a number: Their distance. Instead of starting with such a list, one might also start with a catalog of the convergent sequences or of the open sets. By Theorem 6.21 either one contains all the "topological" information there is in the metric.

156

That is just what we are going to do in Part III, basing ourselves on the catalog of open sets. We will use the opportunity to introduce a wide generalization of our theory.

Inadequacy of Sequences

Examples 10.3
We have set up the theory of metric spaces in order to provide precise formulations for some vague notions of continuity and convergence. If you look back at the examples described in 5.1, you will see that some of them have been incorporated in the theory or in the exercises. The other ones could be treated similarly. One only has to devise suitable metrics.

All the same, one occasionally runs across a type of continuity or convergence that is not generated by a metric. We give two examples.

(i) Let S be a nonempty set and $F(S)$ the set of all real functions on S. In Chapter 8, we have briefly considered pointwise convergence. For $g, f_1, f_2, \ldots \in F(S)$ we say that

$$f_n \to g \text{ pointwise}$$

if

$$f_n(s) \to g(s) \text{ for every } s \in S.$$

This definition yields a very natural mode of convergence in $F(S)$. Is there a metric behind it? Does there exist a metric d on $F(S)$ such that

$$d\text{-convergence} \;=\; \text{pointwise convergence?}$$

It turns out that the answer depends on the size of the set S. If, say, S contains only one element, s_0, then pointwise convergence of $f_1, f_2, \ldots$ to g boils down to $f_n(s_0) \to g(s_0)$ so that we may take the metric

$$d(f, g) := |f(s_0) - g(s_0)| \qquad (f, g \in F(S)).$$

More generally, the answer to our question is affirmative whenever S is finite or even countable (Exercise 10.H). It is not when S is uncountable. Indeed, *if there exists a metric d on $F(S)$ that determines pointwise convergence, then S must be countable.*

Proof
If $s_1, s_2, \ldots$ are pairwise distinct elements of S, then the sequence of indicator functions

$$\mathbf{1}_{\{s_1\}}, \mathbf{1}_{\{s_2\}}, \ldots$$

tends to the zero function $\mathbf{0}$ pointwise (why?), so

$$d(\mathbf{1}_{\{s_n\}}, \mathbf{0}) \longrightarrow 0.$$

Therefore, the set $\{s \in S : d(\mathbf{1}_{\{s\}}, \mathbf{0}) \geq 10^{-1}\}$ cannot contain an infinite sequence of distinct elements, hence it must be finite. In the same way, for every $N \in \mathbb{N}$ the set $S_N := \{s : d(\mathbf{1}_{\{s\}}, \mathbf{0}) \geq N^{-1}\}$ is finite. Then, $S_1 \cup S_2 \cup \ldots$ is countable. (See Theorem 19.8.) But $S_1 \cup S_2 \cup \ldots = \{s : d(\mathbf{1}_{\{s\}}, \mathbf{0}) > 0\} = S$. ∎

(ii) Another example involves continuity.

In Exercise 3.H, we have considered the phenomenon of separate continuity for functions on $\mathbb{R}^2$. We have seen that every continuous function $\mathbb{R}^2 \to \mathbb{R}$ is separately continuous and we have obtained a separately continuous function f_0 that is not continuous.

In contradistinction to separate continuity, the ordinary (Euclidean) continuity is sometimes called *joint continuity*.

Separate continuity is a legitimate object of study, behaving much like the forms of continuity we have considered earlier. For instance, sums and products of separately continuous functions are separately continuous. However, *separate continuity is not generated by any metric*: There is no metric d on $\mathbb{R}^2$ such that for functions on $\mathbb{R}^2$ separate continuity is the same as d-continuity.

Proof

Suppose we have such a metric d. Let us write the elements of $\mathbb{R}^2$ as column vectors. By Example 5.9(iv), the function

$$\begin{pmatrix} x \\ y \end{pmatrix} \longmapsto d\left(\begin{pmatrix} x \\ y \end{pmatrix}, \begin{pmatrix} 0 \\ 0 \end{pmatrix}\right)$$

is d-continuous, hence separately continuous. It follows that

$$d\left(\begin{pmatrix} 1/n \\ 0 \end{pmatrix}, \begin{pmatrix} 0 \\ 0 \end{pmatrix}\right) \longrightarrow 0. \tag{1}$$

By a similar observation, we see that for every n, the function

$$y \longmapsto d\left(\begin{pmatrix} 1/n \\ y \end{pmatrix}, \begin{pmatrix} 1/n \\ 0 \end{pmatrix}\right)$$

is continuous, so that we can choose a number b_n with

$$d\left(\begin{pmatrix} 1/n \\ b_n \end{pmatrix}, \begin{pmatrix} 1/n \\ 0 \end{pmatrix}\right) < n^{-1} \tag{2}$$

$$0 < b_n < n^{-1}. \tag{3}$$

By (1) and (2),

$$d\text{-}\lim_{n\to\infty} \begin{pmatrix} 1/n \\ b_n \end{pmatrix} = \begin{pmatrix} 0 \\ 0 \end{pmatrix}. \tag{4}$$

Now (3) implies that one can make a continuous function $\varphi : \mathbb{R} \to \mathbb{R}$ with $\varphi(n^{-1}) = b_n$ for all n:

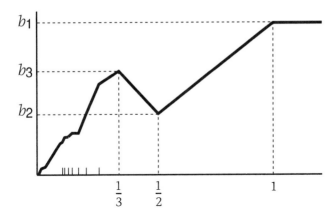

Let $f_0 : \mathbb{R}^2 \to \mathbb{R}$ be the separately but not jointly continuous function mentioned in Exercise 3.H, and define

$$f_1\begin{pmatrix} x \\ y \end{pmatrix} := f_0\begin{pmatrix} \varphi(x) \\ y \end{pmatrix} \qquad \left(\begin{pmatrix} x \\ y \end{pmatrix} \in \mathbb{R}^2 \right).$$

f_1 is separately continuous, hence d-continuous. Then, (4) implies

$$0 = f_1\begin{pmatrix} 0 \\ 0 \end{pmatrix} = \lim_{n \to \infty} f_1\begin{pmatrix} 1/n \\ b_n \end{pmatrix} = \lim_{n \to \infty} f_0\begin{pmatrix} b_n \\ b_n \end{pmatrix} = \frac{1}{2}.$$

Contradiction.   ■

10.4

The contents of 10.1 suggest that we might do well to reconsider the theory of metric spaces and stress the system of all open sets rather than the metric itself. In Example 10.3, we have seen that the whole idea of a metric space may have been too restrictive. In Part III, we will combine these two thoughts and study convergence and continuity in "topological spaces," structures of which metric spaces are particularly simple cases. The remainder of the present chapter is preparative to that. We introduce a more general concept of convergence than we have been using so far, replacing sequences by so-called "nets." It is true that within the context of metric spaces, there is little demand for nets. They do occur now and then (as we will see), but the authors' main purpose in introducing them here is to achieve that in familiar territory the reader can get acquainted with a new tool that farther on will be indispensable.

Examples 10.5

Up to now we have considered limits of *sequences* only. From Analysis, however, we know other types of limits, such as

$$\lim_{x \to 0} \frac{\sin x}{x}, \quad \lim_{x \downarrow 0} e^{-1/x}, \quad \lim_{x \to \infty} e^{-x}, \quad \lim_{|x| \to \infty} \frac{\sin x}{x}.$$

Can they be extended to metric spaces?

The first two do not cause substantial problems. They are special instances of

$$\lim_{\substack{x \to c \\ x \in A}} f(x),$$

where A is a subset of a metric space X, f is a map of A into a metric space X', and $c \in X$. The reader will have little difficulty in inventing a good interpretation. The only trouble is that the value of the limit may not be unique; see Exercise 10.D.

The other two limits,

$$\lim_{x \to \infty} \quad \text{and} \quad \lim_{|x| \to \infty},$$

are not of this type. In the fringes of Analysis there are still other limit-like constructions. We consider two of them.

(i) Let $[a, b]$ be an interval in $\mathbb{R}$. A *partition* of $[a, b]$ is a finite sequence of numbers

$$P = (t_0, t_1, \ldots, t_N)$$

with

$$a = t_0 < t_1 < \cdots < t_N = b.$$

Take a continuous function f on $[a, b]$. For every partition $P = (t_0, t_1, \ldots, t_N)$ of $[a, b]$ we set

$$S_P := f(t_1)(t_1 - t_0) + f(t_2)(t_2 - t_1) + \cdots + f(t_N)(t_N - t_{N-1}).$$

If we let the partition P become finer and finer, so that the t_n come closer and closer together, then the number S_P approaches the integral of f:

$$S_P \longrightarrow \int_a^b f(t)dt.$$

In some sense, the "Riemann sums" S_P converge, but they do not form a sequence. [See Example 10.11(iv) and Exercise 10.F.]

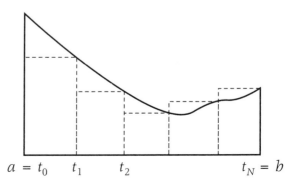

$$a = t_0 \qquad t_1 \qquad t_2 \qquad\qquad\qquad t_N = b$$

(ii) Another instance involves the definition of a sum

$$\sum_{i \in I} t_i,$$

where I is an infinite set (not necessarily $\mathbb{N}$) and for every $i \in I$, t_i is a number. For every finite subset F of I, we consider the "partial sum"

$$S_F := \sum_{i \in F} t_i.$$

The idea now is that $\sum_{i \in I} t_i$ is approximated by S_F if only F is "large enough." The numbers S_F somehow converge, although they do not form a sequence (even if $I = \mathbb{N}$!). [See also Example 10.11(v), and Exercise 10.G.]

For each of the "convergences" we have mentioned, one can devise good and intuitively clear descriptions without going into any new theory of limits. Nevertheless, we prefer to take another point of view. We will define convergence of "nets" and then see how it can be applied in the above cases.

Convergent Nets

10.6
A *direction* in a nonempty set T is a binary relation $\succ$ in T that is *transitive* and *inductive*, i.e.:

D1 If $\tau, \tau', \tau'' \in T$, $\tau \succ \tau'$ and $\tau' \succ \tau''$, then $\tau \succ \tau''$.
D2 For all $\tau_1, \tau_2 \in T$ there is a $\tau \in T$ with $\tau \succ \tau_1$ and $\tau \succ \tau_2$.

Our basic examples are the relations $\geq$ and $>$ in $\mathbb{N}$. For less conventional examples, let T be the collection of all finite subsets of $\mathbb{N}$ and let $\succ$ be $\supset$; or let T be $\mathbb{Z}$ and define $\tau_1 \succ \tau_2$ if and only if τ_1 is a multiple of τ_2.

A *directed set* is a pair $(T, \succ)$ consisting of a nonempty set T and a direction $\succ$ in T.

10.7
Let X be a set. A *net* in X is a map $T \to X$ where $(T, \succ)$ is a directed set. Thus, a sequence of elements of X is a net. Precisely as we have used the notation $(x_n)_{n \in \mathbb{N}}$ for a sequence, we will denote the net

$$\tau \longmapsto x_\tau \qquad (\tau \in T)$$

by

$$(x_\tau)_{\tau \in T}.$$

[See 1.14(x).] The direction is not shown in the notation, but it is an essential part of the definition. Properly speaking, a net in X is a triple consisting of a set T, a direction in T, and a map $T \to X$.

10.8
Let $(T, \succ)$ be a directed set. For all $\tau \in T$, let $P(\tau)$ be some statement or formula. We say that

$$P(\tau) \text{ holds for large } \tau$$

if there exists a $\tau_0 \in T$ such that $P(\tau)$ holds for all $\tau \in T$ with $\tau \succ \tau_0$.

For example, consider the directed set $(\mathbb{N}, \geq)$. Let (X, d) be a metric space, $a, x_1, x_2, \ldots \in X$. Then $x_n \to a$ if and only if

for every $\varepsilon > 0$, we have

$$d(x_n, a) < \varepsilon \text{ for large } n.$$

10.9
Let (X, d) be a metric space, $(x_\tau)_{\tau \in T}$ a net in X, and $a \in X$. We say that the net *converges to a* and that a is a *limit* of the net, and we write

$$x_\tau \longrightarrow a \qquad\qquad (*)$$

if

for every $\varepsilon > 0$, we have

$$d(x_\tau, a) < \varepsilon \text{ for large } \tau.$$

Obviously, convergence of sequences is a special case. For another special case, let T be $(0, \infty)$, directed by the relation $\geq$. If $f : (0, \infty) \to \mathbb{R}$ and $a \in \mathbb{R}$, then the formula

$$f(\tau) \to a$$

as defined in $(*)$, above, means

for every $\varepsilon > 0$, there exists a $\tau_0 > 0$ such that

$$\tau \geq \tau_0 \quad\Longrightarrow\quad |f(\tau) - a| < \varepsilon$$

which in everyday language is just

$$\lim_{\tau \to \infty} f(\tau) = a.$$

Before discussing some examples in detail, we show that a net in a metric space has at most one limit.

Theorem 10.10

Let $(x_\tau)_{\tau \in T}$ be a net in a metric space (X, d). Let $a, b \in X$ and $x_\tau \to a$ and $x_\tau \to b$. Then $a = b$.

Proof

Take $\varepsilon > 0$. There exist τ_1 and τ_2 in T such that

$$\tau \succ \tau_1 \implies d(x_\tau, a) < \varepsilon,$$

$$\tau \succ \tau_2 \implies d(x_\tau, b) < \varepsilon.$$

By Property D2 of $\succ$ (the inductivity) there exists a τ with $\tau \succ \tau_1$ and $\tau \succ \tau_2$. It follows that $d(a, b) < 2\varepsilon$.

This is true for every $\varepsilon > 0$. Hence, $d(a, b) = 0$ and $a = b$. ∎

The theorem allows us to use the notation

$$\lim_\tau x_\tau = a$$

as synonymous to

$$x_\tau \longrightarrow a.$$

Examples 10.11

(i) We have already seen how we obtain the classical

$$\lim_{x \to \infty} f(x)$$

for $f : (0, \infty) \to \mathbb{R}$.

(ii) Take $T := \mathbb{R} \setminus \{0\}$ and define a direction $\succ$ in T by (watch out!)

$$\tau_1 \succ \tau_2 \iff |\tau_1| < |\tau_2|.$$

For a function f on $\mathbb{R} \setminus \{0\}$ and $a \in \mathbb{R}$, we have

$$f(\tau) \to a$$

in the sense of 10.9($*$) if and only if

for every $\varepsilon > 0$, there exists a $\tau_0 \neq 0$ such that

$$\tau \neq 0, |\tau| < |\tau_0| \implies |f(\tau) - a| < \varepsilon$$

which is precisely the

$$\lim_{x \to 0} f(x) = a$$

known from Calculus.

(iii) With $T = \mathbb{R}$ and

$$\tau_1 \succ \tau_2 \iff |\tau_1| > |\tau_2|,$$

we get

$$\lim_{|x|\to\infty} f(x)$$

for $f : \mathbb{R} \to \mathbb{R}$.

(iv) Consider (i) of Example 10.5. Let T be the set of all partitions of $[a, b]$. For partitions $P = (t_0, t_1, \ldots, t_N)$ and $Q = (s_0, s_1, \ldots, s_M)$, one says that P is a *refinement* of Q if $\{t_0, t_1, \ldots, t_N\} \supset \{s_0, s_1, \ldots, s_M\}$; we then write

$$P \succ Q.$$

The relation $\succ$ is a direction in T. For a continuous function f on $[a, b]$, we obtain a net $(S_P)_{P \in T}$ in $\mathbb{R}$. This net can be shown to converge to the integral of f. (See Exercise 10.F.)

(v) (Also, see Exercise 10.G.) For (ii) of Example 10.5, we let T be the collection of all finite subsets of I, directed by $\supset$. This leads to the definition:

$$S = \sum_{i \in I} t_i$$

if

for every $\varepsilon > 0$ there is a finite $F_0 \subset I$ with

$$\left.\begin{array}{l} F \subset I, F \text{ finite} \\[2mm] F \supset F_0 \end{array}\right] \implies \left|\sum_{i \in F} t_i - S\right| < \varepsilon.$$

Convergent nets generalize convergent sequences. Moreover, they occur more or less naturally. Still, the definition would be sterile unless at least part of the theory of sequences can be generalized too. We give three examples of such generalizations; there will be more in the next chapter.

Theorem 10.12
Let X be a metric space; let $Y \subset X$ and $a \in X$. Then a is adherent to Y if and only if there exists a net in Y that converges to a.

Proof
If a is adherent to Y, then we already know that a is the limit of a sequence in Y, and sequences are nets. Conversely, assume there is a net $(x_\tau)_{\tau \in T}$ in Y with $x_\tau \to a$. For every $\varepsilon > 0$, there exists a $\tau_0 \in T$ for which

$$\tau \succ \tau_0 \implies x_\tau \in B_\varepsilon(a);$$

consequently, $B_\varepsilon(a) \cap Y \neq \varnothing$.   ∎

We can also describe continuity in terms of convergent nets.

Theorem 10.13
Let X and X' be metric spaces; let $f : X \to X'$ and $a \in X$. Then, f is continuous at a if and only if

$$x_\tau \to a \text{ in } X \quad \Longrightarrow \quad f(x_\tau) \to f(a) \text{ in } X'. \tag{$*$}$$

Proof
Let d and d' be the metrics in X and X'.

Assume f is continuous at a and let $x_\tau \to a$; we prove $f(x_\tau) \to f(a)$. Let $\varepsilon > 0$. There exists a $\delta > 0$ such that

$$d(x, a) < \delta \quad \Longrightarrow \quad d'\big(f(x), f(a)\big) < \varepsilon.$$

For large τ, we have $d(x_\tau, a) < \delta$, hence $d'\big(f(x_\tau), f(a)\big) < \varepsilon$.

Conversely, if $(*)$ holds, then f certainly is sequentially continuous and therefore continuous at a. ∎

In Theorem 6.23 we have described convergence of a sequence in terms of the system of all open sets. That can also be done for nets. As a consequence, similar metrics allow the same convergent nets.

Theorem 10.14
Let X be a metric space, let $(x_\tau)_{\tau \in T}$ a net in X, and let $a \in X$. Then, $x_\tau \to a$ if and only if

every open set U containing a

contains x_τ for large τ.

Proof
Suppose $x_\tau \to a$ and let U be open, $a \in U$. There is an $\varepsilon > 0$ with $B_\varepsilon(a) \subset U$. For large τ, we have $x_\tau \in B_\varepsilon(a)$, hence $x_\tau \in U$.

For the reverse implication, just note that every ball is open. ∎

Extra: Knots

In Chapter 1 we have seen that a triangle in $\mathbb{R}^2$ can be continuously deformed into a square or a disk. Topologically, these shapes are indistinguishable. In the more formal language we have developed since, we say that the triangle, the square, and the disk are homeomorphic.

This may have left the impression that two homeomorphic subsets of $\mathbb{R}^2$ can always be obtained from each other by a continuous deformation. However, consider the sets

each consisting of a circle and a dot. They certainly are homeomorphic, but we cannot change the first into the second in the sense of Chapter 1—at least as long as we stay within the plane. We could easily do it by going into three-dimensional space.

Given two topological spaces, a quite natural question is whether they are homeomorphic. The question whether they can be deformed into each other is not a priori meaningful; the whole idea of deformation presupposes a fixed surrounding space, such as the plane or three-dimensional space. Given such a "universe," the question makes good sense.

A *knot* is a subset of $\mathbb{R}^3$ that is homeomorphic to a circle, such as a plane circle itself [(i) in the picture, below] or the cloverleaf [(ii) and (iii)]. (Usually one assumes some differentiability, but let us not worry about details.)

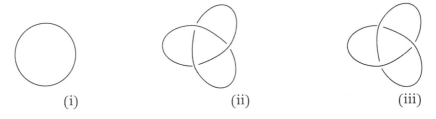

Knot theory deals with deformations of knots. Let us call two knots *equivalent* if within $\mathbb{R}^3$ they can be deformed into each other. The basic question of knot theory is: Given two knots, how can one find out if they are equivalent? (There seems to be some arbitrariness in the choice of $\mathbb{R}^3$ as the surrounding space, but there is a good reason. In $\mathbb{R}^2$, there is not enough room for nontrivial knots at all; in $\mathbb{R}^4$, there is so much room that *all* knots can be obtained by deforming a circle.)

Obviously, the cloverleaf (ii) is equivalent to (iv), but how about (v)?

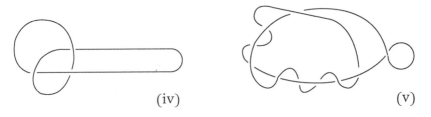

The following sequence illustrates how (ii) can be obtained from (v):

This is not really difficult to do. But how about (i) and (ii): Are they equivalent? A bit of experimenting with a piece of string suggests they are not, but that does not prove anything except, possibly, lack of ingenuity.

Heinrich Tietze (whose name heads Chapter 15) proved in 1908 that the circle (i) and the cloverleaf (ii) are not equivalent. In 1914, Max Dehn showed that the cloverleaves (ii) and (iii) are not equivalent either. [No, you will not get (iii) from (ii) by flipping it over. Try it out and see what happens.]

How does one go about proving that it is *impossible* to change one knot into another by deforming? The mode of attack for Tietze, Dehn, and others since them has been the following. (Sorry, we have to be vague. Details tend to be very technical.) Try to assign to every knot K an integer $n(K)$ in such a way that a small enough deformation will not change the integer. A continuous deformation is a combination of small deformations and therefore will not change the integer either. Hence, if K_1 and K_2 are knots and $n(K_1) \neq n(K_2)$, then K_1 and K_2 will not be equivalent.

Such a function $K \mapsto n(K)$ is called a "knot invariant." In the above, we said that $n(K)$ had to be an integer, but that was only by way of an example. Invariants can also be sets, polynomials, or more abstract objects such as groups. Accordingly, knot theory ties in with diverse topics (e.g., graph theory, group theory, number theory, and functional analysis). Knot theory is an active branch of mathematics with a lot of unsolved problems and, although such an event is rare, as recently as 1984 a new knot invariant was discovered by V. Jones.

Further Reading

Jones, V.F.R., Knot Theory and Statistical Mechanics, *Scientific American*, Nov. 1990, 52-57.

Moran, S., *The Mathematical Theory of Knots and Braids, An Introduction*, North-Holland, Amsterdam, 1983.

Stewart, I., Knots, Links and Videotape, *Scientific American*, Jan. 1994, 134-136.

Exercises

10.A. Under the ordering $\geq$, the interval $[0, 1]$ is a directed set. Let $(x_\tau)_{\tau \in [0,1]}$ be a net in some metric space. Show that the net converges.

(A side comment: Observe that on $[0, 1]$, the relation $>$ is *not* a direction.)

10.B. Let $(S, \succ)$ and $(T, \succ)$ be directed sets. Form the Cartesian product $S \times T$, consisting of all pairs (σ, τ) with $\sigma \in S$, $\tau \in T$. For (σ, τ) and (σ', τ') in $S \times T$, define

$$(\sigma, \tau) \succ (\sigma', \tau') \iff \sigma \succ \sigma' \text{ and } \tau \succ \tau'.$$

Show that in this way $S \times T$ becomes a directed set.

Now suppose $(x_\sigma)_{\sigma \in S}$ and $(y_\tau)_{\tau \in T}$ are nets in $\mathbb{R}$, converging to x and y, respectively. Show that the net $(x_\sigma + y_\tau)_{(\sigma, \tau) \in S \times T}$ converges to $x + y$.

10.C. Let ω be the collection of all open subsets of $\mathbb{R}$ that contain $\sqrt{2}$. Define the relation $\succ$ on ω by

$$U \succ V \iff U \subset V.$$

(We do mean " $\subset$ "!)

Show that $\succ$ is a direction in ω.

Suppose $(x_U)_{U \in \omega}$ is a net in $\mathbb{R}$ with $x_U \in U$ for every U. Show that this net converges to $\sqrt{2}$.

10.D. Let (X, d) and (X', d') be metric spaces. Let $c \in X$, $A := X \setminus \{c\}$.

(i) Let $z \in X'$. Give a sensible definition of the formula

$$\lim_{\substack{x \to c \\ x \in A}} f(x) = z.$$

(ii) Suppose c is an isolated point of X. (See Exercise 7.G.) Show that the above formula holds for every $z \in X'$.

10.E. Let T be the collection of all nonempty bounded subsets of $\mathbb{R}^2$. Define the relation $\succ$ in T by

$$A \succ B \iff A \supset B.$$

(i) Show that $\succ$ is a direction.
(ii) For $A \in T$, let

$$u_A := \inf \left\{ \frac{1}{1 + |x| + |y|} : (x, y) \in A \right\}.$$

Show that the net $(u_A)_{A \in T}$ converges to 0.

(iii) For $A \in T$, let

$$w_A := \inf \left\{ \frac{1}{1 + |x|} : (x, y) \in A \right\}.$$

Does the net $(w_A)_{A \in T}$ converge to 0?

10.F. [Refers to Examples 10.5(i) and 10.11(iv).] Let f be a continuous function on $[a, b]$. Show that the net $(S_P)_{P \in T}$ described in Example 10.11(iv) converges to $\int_a^b f(x)dx$. Hint: Let $\varepsilon > 0$. By (uniform) continuity, there exists a $\delta > 0$ such that

$$x, y \in [a, b], |x - y| \le \delta \implies |f(x) - f(y)| \le \varepsilon.$$

Take any partition $P_0 = (t_0, t_1, \ldots, t_N)$ with $t_n - t_{n-1} \le \delta$ for each n and prove

$$P \succ P_0 \implies |S_P - \int_a^b f(x)dx| \le \varepsilon(b-a).$$

10.G. [Regarding Example 10.11(v).] Let T be the collection of all finite subsets of $\mathbb{N}$. The relation $\supset$ is a direction in T.

Let $t_1, t_2, \ldots$ be a sequence of real numbers. For $F \sqsubseteq T$, put

$$S_F := \sum_{i \in F} t_i.$$

We obtain a net $(S_F)_{F \sqsubseteq T}$ in $\mathbb{R}$. We investigate convergence of this net.

(i) Assume $t_i \ge 0$ for every i, and $\sum_{i=1}^{\infty} t_i = S$. Show that, then, the net converges to S. (Hint: Let $\varepsilon > 0$. There exists an $N \in \mathbb{N}$ with $\sum_{i=1}^{N} t_i > S - \varepsilon$. Note that $\sum_{i=1}^{M} t_i \le S$ for all $M \in \mathbb{N}$.)

(ii) But now consider the case $t_i = (-1)^{i-1} i^{-1}$ ($i \in \mathbb{N}$). You know from Calculus that $\sum_{i=1}^{\infty} t_i$ exists and equals $\ln 2$. Show that, however, the net $(S_F)_{F \sqsubseteq T}$ does not converge. (Hint: Suppose $S_F \to S$ for some $S \in \mathbb{R}$. There is an $F_0 \sqsubseteq T$ such that

$$F \sqsubseteq T, F \supset F_0 \implies |S_F - S| < 1.$$

Deduce that

$$F \sqsubseteq T, F \supset F_0 \implies S_{F \setminus F_0} < 2.$$

Obtain a contradiction from the known fact that $t_1 + t_3 + t_5 + \cdots = \infty$.)

10.H. For any set S, let $F(S)$ be the set of all real-valued functions on S. In Example 10.3(i), we saw that if S is uncountable, then pointwise convergence is not determined by any metric on $F(S)$. In this exercise, we address ourselves to the countable case.

(i) First, let S be a finite set:

$$S = \{s_1, s_2, \ldots, s_N\}$$

with $s_n \ne s_m$ as soon as $n \ne m$. Then $F(S)$ is essentially $\mathbb{R}^N$. Make a metric d on $F(S)$ such that d-convergence is precisely pointwise convergence.

(ii) Now let S be countably infinite:

$$S = \{s_1, s_2, \ldots\}$$

with $s_n \ne s_m$ as soon as $n \ne m$.

Let K be as in Example 9.11:

$$K := \{f \in l^\infty(\mathbb{N}) : 0 \le f(n) \le n^{-1} \text{ for every } n\}.$$

By Exercise 9.G, uniform convergence is the same as pointwise convergence for sequences in K. To every $f \in F(S)$ we assign a function f'

on $\mathbb{N}$:

$$f'(n) := \frac{1}{\pi n}\left(\frac{\pi}{2} + \tan^{-1} f(s_n)\right) \qquad (n \in \mathbb{N}).$$

Show that $f' \in K$ for every $f \in F(S)$, that the formula

$$d(f, g) := \|f' - g'\|_\infty$$

defines a metric d on $F(S)$, and that for sequences in $F(S)$, d-convergence is just pointwise convergence.

10.I. As we have seen in Example 7.12(ii), $1_\mathbb{Q}$ is not the limit of any *sequence* of continuous functions on $\mathbb{R}$. It is, however, the limit of a *net*. In fact, any function $g : \mathbb{R} \to \mathbb{R}$ is the limit (pointwise) of a net of continuous functions. To prove this, take a function g on $\mathbb{R}$.

Let $\mathcal{F}$ be the set of all finite nonempty subsets of $\mathbb{R}$. The relation

$$V \succ W \iff V \supset W$$

is a direction in $\mathcal{F}$. To each $V \in \mathcal{F}$ we assign a continuous function f_V as follows. Let $s_1, s_2, \dots s_N$ be the elements of V, $s_1 < s_2 < \cdots < s_N$. Then, for f_V we take the function with the following properties:

$$\left[\begin{array}{l} \text{its value at } s_n \text{ is } g(s_n) \ (n = 1, \dots, N); \\ \text{on } (-\infty, s_1] \text{ it is constant;} \\ \text{on } [s_N, \infty) \text{ it is constant;} \\ \text{on each interval } [s_n, s_{n+1}] \text{ it is a polynomial of degree } \leq 1. \end{array}\right.$$

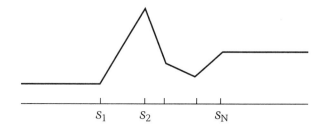

(Do not bother to set up a nice formula defining f_V.)

Show that the net $(f_V)_{V \in \mathcal{F}}$ converges to g pointwise.

11

C H A P T E R

Transition to Topology

We start working on the program that was indicated in the preceding chapter: We wish to reorganize and extend the theory of metric spaces, suppressing the metric and emphasizing the collection of the open sets. There will be much talk of convergence of nets. In case you are still wary of nets, you can mitigate the culture shock by mentally substituting "sequence" for "net." For the time being, the difference will be irrelevant. You will be fairly warned as soon as the nets become essential to the plot.

Generalized Convergence

11.1
Let X be a set and ω any collection of subsets of X, called "lumps."

You may think of X as a metric space, the lumps being the open sets, but X could also be a plane and the lumps the triangles, or X could be $\mathbb{Q}$ and the lumps its finite subsets.

For a net $(x_\tau)_{\tau \in T}$ in X and an element a of X, we define

$$x_\tau \to a$$

if

> every lump U containing a
>
> contains x_τ for large τ.

In such a situation, we will say that the net *converges* to a and that a is a *limit* of the net. Often, at least in the beginning, we will have to indicate the collection of sets we work with and use the terms "ω-convergence" and "ω-limit" and write

$$x_\tau \xrightarrow{\ \omega\ } a.$$

Examples 11.2

(i) If (X, d) is a metric space and the lumps are the d-open sets, then, of course, ω-convergence is precisely d-convergence (Theorem 10.14).

(ii) Again, we let (X, d) be a metric space, but this time the lumps are the d-balls.

If a net $(x_\tau)_{\tau \in T}$ is ω-convergent to a point a, then for every $\varepsilon > 0$ we have

$$x_\tau \in B_\varepsilon(a) \text{ for large } \tau$$

because $B_\varepsilon(a)$ is a lump that contains a. Hence, ω-convergence implies d-convergence. But the converse is also true. Indeed, if $(x_\tau)_{\tau \in T}$ is d-convergent to a and if U is a lump ($=$ a ball) containing a, then $U \supset B_\delta(a)$ for some $\delta > 0$ (Lemma 6.18), so that $x_\tau \in U$ for large τ.

Thus, ω-convergence is the same as d-convergence.

(iii) Let X be any set and let the lumps be the singleton subsets of X. (A *singleton* is a set that has exactly one element.) For a net $(x_\tau)_{\tau \in T}$ in X and an element a of X, we see that

$$x_\tau \xrightarrow{\ \omega\ } a \iff \text{there is a } \tau_0 \in T \text{ such that}$$

$$x_\tau = a \text{ for all } \tau \succ \tau_0.$$

This means that ω-convergence is the same as convergence relative to the trivial metric; see 5.4(vi).

(iv) Let X be any set and let the only lump be X itself. We now come to the disconcerting observation that in this world, *every* element of X is a limit of *every* net in X.

Obviously, this convergence is not generated by any metric (provided that X contains more than only one element) and our new concept of convergence is really more general than the one of Part II. By the same token, one may well ask if we have not overreached ourselves. Is there any sense in a convergence theory with nonunique limits? Remarkably, there is. Such convergences occur naturally in various branches of mathematics, as diverse as abstract algebra and probability theory. (Admittedly, the above example is somewhat extreme.)

(v) One more example to show that we are not just talking general abstract nonsense. It is a bit harder than the previous four. In Example 10.3(i) we have seen that pointwise convergence in the space $F(S)$ of all functions on a set S is not covered by the theory of metric spaces if S is, for instance $[0, 1]$. We will see that our new "theory" is more suitable.

For X, we take, of course, $F(S)$. If U is an open subset of $\mathbb{R}$ and if $s \in S$, we put

$$U_s := \{f \in F(S) : f(s) \in U\}.$$

These sets U_s ($U \subset \mathbb{R}$ open; $s \in S$) are going to be the lumps.

Let $(f_\tau)_{\tau \in T}$ be a net in $F(S)$ and let $g \in F(S)$. We will show that

$$f_\tau \xrightarrow{\omega} g \quad \Longleftrightarrow \quad f_\tau(s) \to g(s) \text{ for every } s \in S.$$

Indeed, we have

$$f_\tau \xrightarrow{\omega} g \quad \Longleftrightarrow \quad \text{if } U_s \text{ is a lump and } g \in U_s,$$
$$\text{then } f_\tau \in U_s \text{ for large } \tau$$
$$\Longleftrightarrow \quad \text{if } s \in S, \text{ if } U \subset \mathbb{R} \text{ is open and } g(s) \in U,$$
$$\text{then } f_\tau(s) \in U \text{ for large } \tau$$
$$\Longleftrightarrow \quad \text{for every } s : \text{ if } U \subset \mathbb{R} \text{ is open and } g(s) \in U,$$
$$\text{then } f_\tau(s) \in U \text{ for large } \tau$$
$$\Longleftrightarrow \quad \text{for every } s : f_\tau(s) \to g(s).$$

11.3
Every collection ω of subsets of X generates a convergence. The following observation is immediate from the definition *if ω_1 is a subcollection of ω_2, then ω_2-convergence implies ω_1-convergence.*

Two distinct collections of subsets may well generate the same convergence. This is illustrated by (i) and (ii) of Example 11.2. Euclidean convergence in $\mathbb{R}$ is generated not only by the collection of all open sets but also by the collection of all balls. In general, when a collection ω of subsets of X and a metric d on X are such that

$$x_\tau \xrightarrow{\omega} x \quad \Longleftrightarrow \quad x_\tau \xrightarrow{d} x,$$

we say that ω and d are *compatible*. Thus, any metric d is compatible with the collection of all d-open sets, and also with the collection of all d-balls.

11.4
Our program for the present chapter is as follows. Let ω be a collection of subsets of a set X. Starting from ω-convergence as defined in 11.1, we introduce notions such as "ω-adherent point," "ω-boundary," "ω-open set," in complete analogy to what we did for metric spaces in Chapter 6. Each time we will see that the new concept coincides with the old in the case when ω is compatible with a metric. Once we have established the new

definitions, we will develop a theory that closely parallels the one we have for metric spaces.

11.5
Let ω be a collection of subsets of X.

Take $A \subset X$. An element c of X is said to be ω-*adherent* to A if there exists a net in A that ω-converges to c.

Every element c of A is ω-adherent to A. (Take any directed set T and put $x_\tau := c$ for all $\tau \in T$.)

By virtue of Theorem 10.12, if ω is compatible with a metric d, then ω-adherence is the same as d-adherence.

11.6
The ω-*boundary* of a set $A \subset X$,

$$\partial A,$$

is the set of all points of X that are ω-adherent to both A and $X \backslash A$. Trivially,

$$\partial A = \partial(X \backslash A).$$

By definition, a subset A of X is ω-*open* if $\partial A \subset X \backslash A$, and ω-*closed* if $\partial A \subset A$.

From the closing lines of 11.5, we see that if ω is compatible with a metric d, then the ω-boundary is the d-boundary, ω-openness is the same as d-openness, and ω-closedness is the same as d-closedness.

Theorem 11.7
If A and B are subsets of X that are each other's complements, then

$$A \text{ is } \omega\text{-closed} \iff B \text{ is } \omega\text{-open}.$$

Proof
$\partial A = \partial B.$ ∎

Theorem 11.8
Let ω be a collection of subsets of X. Let $A \subset X$.

(i) *The set of points ω-adherent to A is precisely $A \cup \partial A$.*
(ii) *A is ω-closed if and only if every point ω-adherent to A is an element of A.*

Proof
If c is ω-adherent to A but $c \notin A$, then $c \in X \backslash A$, so c is ω-adherent to $X \backslash A$, from which $c \in \partial A$. This essentially proves (i). Part (ii) is an immediate consequence. ∎

The ω-open sets form a collection ω^*.

Lemma 11.9

(i) $\omega \sqsubset \omega^*$, i.e., *every set belonging to ω is ω-open.*
(ii) *ω-convergence and ω^*-convergence are the same.*

Proof

(i) Let $U \in \omega$. *Suppose $U \cap \partial U$ contains a point c. As c is ω-adherent to $X \backslash U$, there is a net $(x_\tau)_{\tau \in T}$ in $X \backslash U$ that ω-converges to c. But, as $c \in U$, by definition of ω-convergence this means that $x_\tau \in U$ for large τ. Contradiction.*

(ii) It follows from (i) that ω^*-convergence entails ω-convergence. Conversely, let $(x_\tau)_{\tau \in T}$ be a net in X that ω-converges to $c \in X$; we show that it ω^*-converges to c. Take $W \in \omega^*$ and $c \in W$. We wish to prove that there is a $\tau \in T$ such that

$$\sigma \in T, \ \sigma \succ \tau \implies x_\sigma \in W.$$

Suppose no such τ exists; i.e., suppose that for every $\tau \in T$ we have a $\tau' \in T$ with

$$\tau' \succ \tau \quad \text{but} \quad x_{\tau'} \notin W.$$

Thus, we obtain a net $(x_{\tau'})_{\tau \in T}$ in $X \backslash W$. This net is easily seen to ω-converge to c. Indeed, take $U \in \omega$ and $c \in U$. There is a $\tau_0 \in T$ with

$$\tau \in T, \ \tau \succ \tau_0 \implies x_\tau \in U.$$

Then,

$$\tau \in T, \ \tau \succ \tau_0 \implies \tau' \succ \tau_0 \implies x_{\tau'} \in U.$$

Hence, $x_{\tau'} \overset{\omega}{\longrightarrow} c$. Then, c is ω-adherent to $X \backslash W$. But $X \backslash W$ is ω-closed (Theorem 11.7), so $c \in X \backslash W$ [Theorem 11.8(ii)]. *Contradiction.* ∎

11.10

(i) It follows that ω^* is the largest collection of sets that generates ω-convergence. Indeed, if φ is any collection of sets such that

$$x_\tau \overset{\varphi}{\longrightarrow} x \iff x_\tau \overset{\omega}{\longrightarrow} x,$$

then the φ-open sets are just the ω-open sets, i.e., $\varphi^* = \omega^*$. But $\varphi \sqsubset \varphi^*$, so $\varphi \sqsubset \omega^*$.

(ii) By (ii) of 11.6, ω-openness is the same as ω^*-openness, so $\omega^* = (\omega^*)^*$.

Theorem 6.15 has a straightforward extension:

Theorem 11.11

(I)   (i) *The intersection of any two ω-open subsets of X is ω-open.*

   (ii) *The union of any nonempty collection of ω-open subsets of X is ω-open.*

(II)   (i) *The union of any two ω-closed subsets of X is ω-closed.*

   (ii) *The intersection of any nonempty collection of ω-closed subsets of X is ω-closed.*

Proof

We start with (I)(i). Let U and V be ω-open.

Suppose $U \cap V$ contains a point c of $\partial(U \cap V)$. As c is ω-adherent to $X \backslash (U \cap V)$, there exists a net $(x_\tau)_{\tau \in T}$ in $X \backslash U \cap V$ that ω-converges to c. But ω-convergence is ω^*-convergence, and U and V belong to ω^* and both contain c. Therefore, there exist $\tau_U, \tau_V \in T$ with

$$\tau \in T, \ \tau \succ \tau_U \implies x_\tau \in U,$$

$$\tau \in T, \ \tau \succ \tau_V \implies x_\tau \in V.$$

Taking any $\tau \in T$ with $\tau \succ \tau_U$ and $\tau \succ \tau_V$, we obtain $x_\tau \in U \cap V$. *Contradiction.*

This proves (I)(i). Now (II)(i) is a simple consequence: If A and B are ω-closed, then $X \backslash A$ and $X \backslash B$ are ω-open; then so is their intersection, which is $X \backslash (A \cup B)$; then $A \cup B$ is ω-closed.

By a similar technique, (I)(ii) will follow from (II)(ii), so it suffices to prove (II)(ii). That is easy to do. Let α be any nonempty collection of ω-closed sets. Let c be ω-adherent to $\bigcap \alpha$; by Theorem 11.8(ii), we are done if we prove $c \in \bigcap \alpha$. Now for every $A \in \alpha$, c is ω-adherent to A (since $\bigcap \alpha \subset A$), so $c \in A$, again by Theorem 11.8(ii). Hence, $c \in \bigcap \alpha$. ∎

11.12

A collection φ of subsets of X is called a *topology on X* if

> **T1** The intersection of any two sets belonging to φ again belongs to φ.
> **T2** The union of any nonempty subcollection of φ belongs to φ.
> **T3** $\varnothing$ and X belong to φ.

It is easy to see that $\varnothing$ and X always are ω-open. Thus:

Theorem 11.13

For any collection ω of subsets of X the ω-open sets form a topology in X.

11.14

Next, we consider continuity. Let ω be a collection of subsets of a set X and ω' a collection of subsets of a set X'. Let $f : X \to X'$.

We say f is *ω-ω'-continuous at a* if

$$x_\tau \xrightarrow{\omega} a \text{ in } X \implies f(x_\tau) \xrightarrow{\omega'} f(a) \text{ in } X'.$$

From Theorem 10.13, we see that if X and X' are metric spaces and ω and ω' determine the metric convergences, then ω-ω'-continuity is just the familiar metric continuity.

Of course, we say that f is *ω-ω'-continuous* if it is ω-ω'-continuous at every point of X.

The generalization of Theorem 6.24 proceeds without trouble:

Theorem 11.15
Let X,ω,X', and ω' be as above. For $f:X \to X'$, the following conditions (α), (β), and (γ) are equivalent:
(α) f is continuous.
(β) For every ω'-open set $V \subset X'$, its inverse image $f^{-1}(V)$ is ω-open.
(γ) For every ω'-closed set $A \subset X'$, its inverse image $f^{-1}(A)$ is ω-closed.

Proof
(α) $\implies$ (γ) Let $A \subset X'$ be ω'-closed. Let $a \in X$ be ω-adherent to $f^{-1}(A)$; we prove $a \in f^{-1}(A)$. There is a net $(x_\tau)_{\tau \in T}$ in $f^{-1}(A)$ with $x_\tau \to a$. Then $f(x_\tau) \to f(a)$ in X' and $f(x_\tau) \in A$ for every τ, so $f(a) \in A$ and $a \in f^{-1}(A)$.
(β) $\implies$ (α) Let $x_\tau \to x$ in X. To prove that $f(x_\tau) \to f(x)$ in X', take an ω'-open $V \subset X'$ with $f(x) \in V$. Then, $f^{-1}(V)$ is ω-open and $x \in f^{-1}(V)$. Hence, for large τ, we have $x_\tau \in f^{-1}(V)$ and $f(x_\tau) \in V$.
(β) $\iff$ (γ) is left to the reader. ∎

11.16
Following 6.25, we define the *ω-interior* of a subset A of X,

$$\text{int}(A) \quad \text{or} \quad A^\circ,$$

to be the union of all ω-open sets that are contained in A. It is itself ω-open and, therefore, is the largest ω-open set contained in A.

Also, the *ω-closure* of A,

$$\text{clo}(A) \quad \text{or} \quad A^-,$$

is the intersection of all ω-closed sets that contain A; it is the smallest ω-closed set containing A.

11.17
In the case when ω is compatible with a metric d, the ω-closure and the ω-interior are just the d-closure and the d-interior, respectively.

For this situation, Theorem 6.26 gives us an alternative description of A^-: It is precisely the set of all points adherent to A. The same is the case in our present, more general situation, as we will prove in Lemma 11.18.

First, an observation on sequences and nets. So far in this chapter, we could have done everything with converging *sequences* without ever mentioning nets. We could have introduced sequentially adherent point and sequentially open and closed sets. The sequentially open sets form a topology, and it makes sense to define the sequential closure. It is, however, not generally true that elements of the sequential closure of A are sequentially adherent to A; we present a counterexample in Example 11.27.

Thus, for a good extension of Theorem 6.26, sequences are insufficient. Nets will do, as we proceed to show.

Lemma 11.18
Let ω be a collection of subsets of X. For $A \subset X$ and $c \in X$, the following conditions are equivalent:

(α) $c \in A^-$; i.e., c is an element of every ω-open set containing A.
(β) c is ω-adherent to A.
(γ) A intersects every ω-open set that contains c.

Proof
The equivalence of (α) and (γ) is a beginner's exercise in set theory. The implication (β) $\implies$ (α) follows from Theorem 11.8(ii). We prove (γ) $\implies$ (β), and this is where the nets come in.

Assume (γ). We construct a net in A that ω-converges to c. (Fair warning: The underlying directed set is going to be a bit strange.)

The ω-open sets containing c form a collection ω_c. If two sets both belong to ω_c, then so does their intersection. It follows that we can define a direction $\succ$ in ω_c by

$$W_1 \succ W_2 \iff W_1 \subset W_2.$$

(Note the direction of the inclusion symbol $\subset$.) For every $W \in \omega_c$, by (γ) we can choose a point x_W with

$$x_W \in W \cap A.$$

In this way, we have obtained a net $(x_W)_{W \in \omega_c}$ in A. We show that this net ω-converges to c. That is very easy to do: Let $W_0 \in \omega$ and $c \in W_0$. Then, $W_0 \in \omega_c$. For every $W \in \omega_c$ with $W \succ W_0$, we have $W \subset W_0$ and therefore $x_W \in W \subset W_0$. Hence,

$$W \in \omega, \; W \succ W_0 \implies x_W \in W_0,$$

and we are done. ∎

Topologies

11.19
We have essentially completed our program of extending the theory of metric spaces. There still remain things to be said about, say, interior points and dense sets, but they are more or less routine exercises.

(See Chapter 12.) Compactness is a different matter; we treat it more extensively in Chapter 13.

In the balance of the present chapter, we elaborate on the relation between ω and ω^*, the topology of all ω-open sets. In particular, we describe a simple way to construct ω^* directly from ω without intervention of convergence. We will also see that ω^* is equal to ω as soon as ω is a topology.

We look back at our starting-point: metric spaces. Let d be a metric on X and let ω be the collection of all balls. Then ω-convergence is the same as d-convergence [Example 11.2(ii)], so ω-openness is d-openness and ω^* is the collection of all d-open sets. The d-open sets are precisely the unions of balls (Theorem 6.19). Thus, in this situation, for $U \subset X$ we have

$$U \in \omega^* \iff U \text{ is the union of} \qquad (*)$$

$$\text{a subcollection of } \omega.$$

The relation between ω and ω^* is not always as simple as that. For instance, let X be $\mathbb{R}$ and let ω be the collection of all intervals of length 2. The intervals $(0, 2)$ and $(1, 3)$ belong to ω and therefore are ω-open; then, so is their intersection. Thus, $(1, 2) \in \omega^*$, but, of course, $(1, 2)$ is not a union of intervals of length 2. Still, the situation $(*)$ occurs often enough to merit special attention.

11.20

Let φ and ω be collections of subsets of X. We say that ω is a *base* for φ if

$$U \in \varphi \iff U \text{ is the union of}$$

$$\text{a subcollection of } \omega.$$

Thus, in a metric space, the balls form a base for the topology of all open sets. (The open sets themselves form another one.)

11.21

For a less formal language, we fall back on our earlier terminology and call the sets belonging to ω "lumps," again. Then, ω is a base for φ if

$$U \in \varphi \iff U \text{ is a union of lumps.}$$

In practice, how do we go about to prove that a given set U is a union of lumps? On the one hand, if it is, then for every point a of U, there exists a lump W with

$$a \in W \quad \text{and} \quad W \subset U.$$

On the other hand, if for every $a \in U$, we can find such a lump W, then U is a union of lumps (e.g., U is the union of all lumps that are contained in U).

Apart from the case of metric spaces, we have the following situation in which (∗) of 11.19 is pertinent.

Lemma 11.22

Let ω be a collection of subsets of X such that

$$X \in \omega, \tag{1}$$

$$W_1, W_2 \in \omega \implies W_1 \cap W_2 \in \omega. \tag{2}$$

Then, ω is a base for the topology of all ω-open sets.

Proof

We use the terminology of the lumps.

On the one hand, every union of lumps is a union of ω-open sets [Lemma 11.9(i)], hence it is itself ω-open. Conversely, let U be an ω-open set; we wish to prove that U is a union of lumps. Take $a \in U$; in view of 11.21, we are done if there exists a lump W with

$$a \in W \quad \text{and} \quad W \subset U.$$

(The following reasoning is quite analogous to our proof of Lemma 11.18.)

The lumps containing a form a collection ω_a:

$$W \in \omega_a \iff W \text{ is a lump, } a \in W.$$

By (1), this collection is not empty; by (2), it is directed if we define

$$W_1 \succ W_2 \iff W_1 \subset W_2.$$

For every $W \in \omega_a$, choose a point x_W of W, subject only to the restriction that, if possible, x_W lies in $X \backslash U$:

$$x_W \in W,$$

$$x_W \in X \backslash U \text{ if } W \text{ intersects } X \backslash U.$$

The points x_W form a net in X. This net ω-converges to a. (Indeed, if $V \in \omega$ and $a \in V$, then $V \in \omega_a$, and $x_W \in V$ as soon as $W \succ V$.) As U is ω-open, it follows that $x_W \in U$ for large W. ("Large" in the sense of $\succ$!) Take any W with $x_W \in U$. Then, we do *not* have $x_W \in X \backslash U$, so W does *not* intersect $X \backslash U$. But then, $W \subset U$ and we are done. ∎

Corollary 11.23

If ω is a topology in X, then ω consists precisely of the ω-open sets.

For any collection ω of subsets of X, we can construct ω^* as follows:

Theorem 11.24

Let ω be a collection of subsets of X. Form the collection $\omega^\cap$ by

$$U \in \omega^\cap \iff \begin{cases} \text{either } U = X \\ \text{or } U = W_1 \cap \ldots \cap W_N \text{ for certain } W_1, \ldots, W_N \in \omega. \end{cases}$$

Then, $\omega^\cap$ is a base for the topology of all ω-open sets.

Proof

$\omega \sqsubset \omega^\cap \sqsubset \omega^*$, so $\omega^* \sqsubset \omega^{\cap*} \sqsubset \omega^{**}$; but $\omega^{**} = \omega^*$. Hence, $\omega^* = \omega^{\cap*}$.

Thus, ω-convergence and $\omega^\cap$-convergence are the same. Then, the ω-open sets are precisely the $\omega^\cap$-open sets. Now clearly,

$$X \in \omega^\cap$$

$$U_1, U_2 \in \omega^\cap \implies U_1 \cap U_2 \in \omega^\cap.$$

By Lemma 11.22, $\omega^\cap$ is a base for the topology of the $\omega^\cap$-open (= ω-open) sets. ∎

The above gives a concrete way *to build ω^* out of ω: Form finite intersections; add X; form arbitrary unions*. There is also a more abstract description of ω^*:

Theorem 11.25

Let ω be a collection of subsets of X. Then, ω^* is the smallest topology containing ω, That is,

(i) *ω^* is a topology; $\omega \sqsubset \omega^*$;*
(ii) *if φ is a topology and $\omega \sqsubset \varphi$, then $\omega^* \sqsubset \varphi$.*

Proof

For (i), see Theorem 11.13 and Lemma 11.9(i). For (ii), if φ is a topology and $\omega \sqsubset \varphi$, then $\omega^\cap \sqsubset \varphi$, hence (by Theorem 11.24) $\omega^* \sqsubset \varphi$. ∎

11.26

If ψ is a topology in X and ω is a collection of subsets of X, one says that ω is a *subbase* for ψ (or that ψ is the topology *generated* by ω) if ψ is the smallest topology containing ω. Apparently, this is the case if and only if $\psi = \omega^*$; i.e., ψ consists of the ω-open sets.

Every collection ω of subsets of X is a subbase for a topology (constructible by forming finite intersections, adding X, and then forming arbitrary unions).

Every base for a topology ψ is also a subbase for ψ. (Why?) Every topology is a (sub)base for itself.

Example 11.27

We owe the reader an example showing that sequences are not enough to have a good generalization of Theorem 6.26. (See 11.17.) In this section we present a set X with a collection ω of "lumps" (not at all artificial) and a subset A such that the set A^s of all points that are sequentially adherent to A is not sequentially closed (and therefore cannot be the sequential closure of A).

For X, we take $F(\mathbb{R})$, the set of all functions on $\mathbb{R}$. We let ω be as in Example 11.2(v), so that ω-convergence is the familiar pointwise convergence. Let A be the set of all continuous functions. Choose an enumeration $\{q_1, q_2, \ldots\}$ of $\mathbb{Q}$ and let x_n be the indicator of the finite set $\{q_1, \ldots, q_n\}$. As

$$x_n(t) = \lim_{k \to \infty} e^{-k|t-q_1|} + \cdots + e^{-k|t-q_n|} \qquad (t \in \mathbb{R};\ n \in \mathbb{N}),$$

each x_n lies in A^s. Furthermore,

$$x_n \to \mathbf{1}_{\mathbb{Q}} \quad \text{pointwise}.$$

But $\mathbf{1}_{\mathbb{Q}} \notin A^s$, as we have seen in Example 7.12(ii), so A^s is not sequentially closed.

Extra: The Emergence of the Professional Mathematician

King Ptolemy I, in what is called the Golden Age of Greek Mathematics, paid the greatest minds of his days to live, teach, and do research in the so-called Museum of Alexandria. The Marquess de l'Hôpital (1661-1704) could easily afford private lessons in mathematics from Johann Bernoulli (1667-1748). He even bought "his" theorem (l'Hôpital's Theorem) from Bernoulli. Euler was paid by Frederick the Great and Catherine the Great in Berlin and St. Petersburg, respectively. Cauchy, while in Prague in exile from French turmoil, tutored the son of Charles X. Who is paying mathematicians and for what?

The slow rise of universities in Western Europe in the late Middle Ages did not immediately lead to separate mathematics departments. It is interesting that the first mathematicians who pointed in the direction of topology, the Germans Gauss, Listing, and Moebius, all lived in a time when universities changed radically. Both Moebius and Listing were students of Gauss. Gauss and Moebius held positions as astronomers at the universities of Goettingen and Leipzig, respectively, while Listing became professor of Physics in Goettingen.

Moebius was only 16 years old when a few miles away from his hometown the Prussians were crushed by Napoleon's army in the battle of Jena (1806). The defeated Prussians became part of a cultural revolution that is nowadays described as neohumanism. Universities that formerly had few departments started to develop Chairs for new subjects. Botany, chemistry, and mathematics became entities of their own. University professors were required to organize seminars in their specialty serving the purpose of both research activity as well as a chance for students to get to the front of current knowledge. This change in organization of universities turned out to be very fortuitous. As the arts started to blossom with Beethoven, Goethe, Schiller, and others, so did mathematics. Gauss, Moebius, and Listing saw these changes happen during their lifetimes as mathematicians in a nonmathematics university chair. The effects would become obvious in the next generation. Berlin has been among the first to make Professors serve this dual role of researcher and teacher and Berlin became the center of the mathematical universe. Weierstrass and his students investigated the foundations for analysis at the same time as others pursued ideas from the previous generation that would lead to Topology. With mathematics recognized as a separate discipline, a multitude of new directions was soon to be pursued. This development has continued unabated into this century and has led to another Golden Age for mathematics. There is no need to predict the future.

Further Reading

Fauvel, J., R. Flood and R. Wilson, *Möbius and his Band*, Oxford University Press, Oxford, 1993.

Exercises

11.A. A *pseudometric* on a set X is a function d on $X \times X$ satisfying Axioms M1, M3, and M4 mentioned in 5.2, but possibly not, Axiom M2. It still follows that $d(x, x) = 0$ for all x. For instance, let X be $C[0, 1]$ and

$$d(f, g) := |f(0) - g(0)|$$

or let X be the set of all integrable functions on $[0, 1]$ and

$$d(f, g) := \int_0^1 |f(x) - g(x)| dx.$$

If d is a pseudometric on X, one can define d-balls and d-open sets precisely as we did for a metric, and one easily sees that every d-ball is d-open.

(i) Now let $d_1, d_2, \ldots$ be pseudometrics on some set X. Let ω be the collection of all subsets U of X with the property

there is a k such that U is d_k-open.

Show that for a net $(x_\tau)_{\tau\in T}$ in X and $x \in X$

$$x_\tau \xrightarrow{\omega} x \iff d_k(x_\tau, x) \to 0 \text{ for every } k.$$

(ii) An example. Let $C(\mathbb{R})$ be the set of all continuous functions on $\mathbb{R}$. A sequence $f_1, f_2, \ldots$ in $C(\mathbb{R})$ is said to converge *locally uniformly* to $f \in C(\mathbb{R})$ if for every closed bounded interval $[a, b]$

$$f_n \to f \text{ uniformly on } [a, b].$$

For $k \in \mathbb{N}$ and $f, g \in C(\mathbb{R})$, put

$$d_k(f, g) := \|(f - g)|_{[-k,k]}\|_\infty.$$

Show that each d_k is a pseudometric, and that, if we take ω as in (i),

$$f_n \xrightarrow{\omega} f \iff f_n \to f \text{ locally uniformly.}$$

11.B. Let $F(\mathbb{R})$ be the set of all functions on $\mathbb{R}$. For $Z \subset \mathbb{R}$ and $f \in F(\mathbb{R})$, put

$$B_Z(f) := \{g \in F(\mathbb{R}) : f|_Z = g|_Z\}.$$

Observe: if $Z_1, Z_2, \ldots \subset \mathbb{R}$ and $Z = \bigcup_{n\in\mathbb{N}} Z_n$, then for all f

$$B_Z(f) = \bigcap_{n\in\mathbb{N}} B_{Z_n}(f).$$

Let ω be the collection of all subsets U of $F(\mathbb{R})$ with the property

for every $f \in U$, there exists a

countable $Z \subset \mathbb{R}$ with $B_Z(f) \subset U$.

(i) Prove that ω is a topology on $F(\mathbb{R})$.
(ii) Prove: If $U_1, U_2, \ldots \in \omega$, then $\bigcap_{n\in\mathbb{N}} U_n \in \omega$.
(iii) Prove: If $f, f_1, f_2, \ldots \in F(\mathbb{R})$ and $f_n \xrightarrow{\omega} f$, then $f \in \{f_1, f_2, \ldots\}$. (Hint: Otherwise there would exist $x_1, x_2, \ldots \in \mathbb{R}$ with $f(x_1) \neq f_1(x_1)$, $f(x_2) \neq f_2(x_2)$, and so forth Take $Z := \{x_1, x_2, \ldots\}$.)
(iv) Now show that, relative to ω, every subset of $F(\mathbb{R})$ is sequentially closed, but not every subset is closed. (Singleton sets are not open.)

III

PART

Topological Spaces

12

CHAPTER

Topological Spaces

In Chapter 11, we have completed the transition from metric spaces to topological spaces. Starting from a set X and a collection ω of lumps in X, we have introduced notions such as "ω-convergent" and "ω-open." The collection of all ω-open sets is a topology (Theorem 11.13). Then, in Corollary 11.23 we saw that a topology is a very natural collection of lumps to take for ω.

Accordingly, in the remainder of this book, the focus will be on topologies.

One can (and most textbooks do) take topologies as the starting point instead of convergence. To underscore that position, we open this chapter with repeating the definition of "topology." Subsequently, we present a synopsis of the previous chapter, but this time proceeding from the notion of a topology.

If you have read Chapter 11, the sections 12.1 through 12.14 will have nothing new to offer beyond a few minor items of terminology. Have a glance at the definitions of "topological space" (12.1) and "neighborhood" (12.9), consider the examples in 12.2, and resume reading at Example 12.15. In case you do wish to read the intervening sections, you should know in advance that the definitions given there may look quite different from those of Chapter 11 but are equivalent to them.

If you have skipped the previous chapter, Chapter 12 will be your introduction to the subject of topological spaces, self-contained except for some proofs for which due reference will be given.

12.1

A collection ω of subsets of a given set X is called a *topology* on X if it satisfies the requirements:

T1 The intersection of any two sets belonging to ω again belongs to ω.
T2 The union of any nonempty subcollection of ω belongs to ω.
T3 $\varnothing$ and X belong to ω.

A *topological space* is a pair (X, ω) consisting of a set X and a topology ω on X. Often, we will not bother to mention the topology and just say "the topological space X" instead of "the topological space (X, ω)." Confusion will never arise.

Examples 12.2

(i) If d is a metric on X, the d-open sets form a topology, the so-called *d-topology*. [See Theorem 6.15(I).] This example is fundamental for the following theory. A topology ω [or the topological space (X, ω)] is called *metrizable* if there exists a metric d such that ω is the d-topology.

   If X is a subset of $\mathbb{R}^N$ and d_E is the Euclidean metric on X, then the topology of all d_E-open subsets of X is called the *Euclidean topology* on X.

(ii) If X is any set, the collection of all subsets of X is a topology on X, the *discrete topology*. It is metrizable. [See Example 6.9(iii).]

(iii) For any set X, the collection of subsets whose only members are $\varnothing$ and X is the *indiscrete* or *trivial* topology. It is not metrizable if X contains at least two elements.

(iv) Take a set X and let ω consist of

   the empty set $\varnothing$ and

   the complements of the finite subsets of X.

   Then, ω is a topology. It is the discrete topology in case X itself is a finite set.

(v) Let X be $\mathbb{R}$ and let ω consist of

   the empty set,

   the entire space $\mathbb{R}$, and

   all intervals (a, ∞) with $a \in \mathbb{R}$.

   Then, ω is a topology.

(vi) Let X be $\mathbb{R}$. The subsets U of $\mathbb{R}$ with the property

> for every $a \in U$, there exists
>
> an $\varepsilon > 0$ for which $[a, a+\varepsilon) \subset U$

form a topology ω_r. Every set that is open relative to the Euclidean metric belongs to ω_r; so does the interval $[0, 1)$.

12.3
Let (X, ω) be a topological space. A subset of X is *open* if it belongs to ω; it is *closed* if its complement belongs to ω. (In the literature, these are the standard definitions of "open" and "closed"; in Chapter 11 the statements are contained in Corollary 11.23 and Theorem 11.7.) The union of two closed sets is always closed; any intersection of closed sets is closed.

12.4
We make two observations concerning these definitions and, indeed, most definitions in this chapter. First, we do not keep mentioning the topology. We will not say "ω-open," "ω-closed," and so on unless we must. Second, most terms we are going to introduce extend terminology we already have at our disposal for metric spaces. If d is a metric on X and ω is the d-topology, then the words "open" and "closed" as defined above mean precisely "d-open" and "d-closed." You may expect a similar consistency in the rest of this chapter.

12.5
Let (X, ω) be a topological space. If $(x_\tau)_{\tau \in T}$ is a net in X and $a \in X$, the net is said to *converge* to a,

$$x_\tau \longrightarrow a,$$

and a is called a *limit* of the net if

> for every $U \in \omega$ with $a \in U$,
>
> there exists a τ_0 in T such that
>
> $\tau \in T, \ \tau \succ \tau_0 \implies x_\tau \in U.$

An element a of X is *adherent* to a subset Y of X if there is a net in Y that converges to a. (See Theorem 10.12 for the metrizable case.)

Lemma 12.6
Let (X, ω) be a topological space and let $Y \subset X$ and $a \in X$. The following are equivalent:

(α) *a is adherent to Y.*
(β) *a is an element of every closed set that contains Y.*
(γ) *Whenever $W \subset X$ is open and $a \in W$, then $W \cap Y \neq \varnothing$.*

Proof
See Lemma 11.18.   ■

12.7
Let Y be a subset of a topological space X.
    The *closure* of Y,

$$\mathrm{clo}(Y), \quad \text{or} \quad Y^-,$$

is the intersection of all closed sets that contain Y; it is the smallest closed
set containing Y. The *interior* of Y,

$$\mathrm{int}(Y), \quad \text{or} \quad Y^\circ,$$

is the union of all open sets that are contained in Y; it is the largest open
set contained in Y. (See 6.25 for the metrizable case.)
    Trivially,

$$Y^\circ \subset Y \subset Y^-;$$

$$Y \text{ is closed} \iff Y = Y^- \iff Y \supset Y^-;$$

$$Y \text{ is open} \iff Y = Y^\circ \iff Y \subset Y^\circ.$$

It follows from Lemma 12.6 that Y^- *is precisely the set of all points of X that
are adherent to Y.* Hence, Y is closed if and only if every point adherent
to Y actually is an element of Y.
    An important conclusion is that the convergence determines the
topology:

Corollary 12.8
If ω and ω' are topologies on X such that

$$x_\tau \xrightarrow{\omega} a \iff x_\tau \xrightarrow{\omega'} a$$

for all nets $(x_\tau)_{\tau \in T}$ in X and all $a \in X$, then $\omega = \omega'$.

Proof
By the last line of 12.7, the ω-closed sets are precisely the ω'-closed sets,
so the ω-open sets are precisely the ω'-open sets.   ■

12.9
Let Y be a subset of a topological space X.
    We say that a point a of X is *interior* to Y, or that Y is a *neighborhood*
of a if there exists an open set U with $a \in U \subset Y$. Clearly, this is the
case if and only if $a \in Y^\circ$. Thus, Y is open if and only if every point of Y
is interior to Y, i.e., if and only if Y is a neighborhood of every element
of Y.

12.10

Let Y be a subset of a topological space X.

The *boundary* of Y,

$$\partial Y,$$

is the set of all points of X that are adherent to both Y and $X\backslash Y$. In other words,

$$\partial Y := Y^- \cap (X\backslash Y)^-.$$

From 12.7, it is clear that $(X\backslash Y)^- = X\backslash Y^\circ$, so

$$\partial Y = Y^-\backslash Y^\circ.$$

We have

$$\partial Y = \partial(X\backslash Y),$$

$$Y \text{ is open} \quad\Longleftrightarrow\quad Y \cap \partial Y = \varnothing,$$

$$Y \text{ is closed} \quad\Longleftrightarrow\quad \partial Y \subset Y.$$

The sets Y°, $(X\backslash Y)^\circ$, and ∂Y are pairwise disjoint; their union is the whole space X.

12.11

(See 11.20.) A collection φ of subsets of X is a *base* for a topology ω if

$$Y \in \omega \quad\Longleftrightarrow\quad Y \text{ is the union of}$$
$$\text{some subcollection of } \varphi.$$

Equivalently:

$$Y \in \omega \quad\Longleftrightarrow\quad \text{for every } a \in Y \text{ there is}$$
$$\text{a } W \in \varphi \text{ with } a \in W \subset Y.$$

An example: If d is a metric on X, the d-balls form a base for the d-topology. (Theorem 6.19.)

12.12

(See 11.26.) Let λ be any collection of subsets of X. Form a collection ω of subsets of X by

$$Y \in \omega \quad\Longleftrightarrow\quad Y \in \varphi \text{ for every topology } \varphi$$
$$\text{for which } \lambda \sqsubset \varphi.$$

It is straighforward to verify that ω itself is a topology. (For example, let $Y_1, Y_2 \in \omega$. If φ is any topology with $\lambda \sqsubset \varphi$, then $Y_1 \in \varphi$ and $Y_2 \in \varphi$, so $Y_1 \cap Y_2 \in \varphi$. Hence, $Y_1 \cap Y_2 \in \omega$.)

This ω is the smallest topology containing λ; that is

(a) ω is a topology; $\lambda \sqsubset \omega$;
(b) if φ is a topology and $\lambda \sqsubset \varphi$, then $\omega \sqsubset \varphi$.

We say that λ is a *subbase* for ω, or that ω is the topology *generated* by λ. Observe that every base for ω is a subbase.

An explicit construction of ω out of λ is described in:

Theorem 12.13

Let λ be a subbase for a topology ω on a set X. Form the collection $\lambda^\cap$ consisting of the sets

$$\left[\begin{array}{l} X, \\ W_1 \cap \ldots \cap W_N \text{ where } N \in \mathbb{N} \text{ and } W_1, \ldots, W_N \Subset \lambda. \end{array} \right.$$

Then $\lambda^\cap$ is a base for ω.

Proof

Define a collection ω' of sets by

$$Y \Subset \omega' \iff Y \text{ is the union of}$$

$$\text{some subcollection of } \lambda^\cap,$$

i.e.,

$$Y \Subset \omega' \iff \text{for every } a \in Y \text{ there is}$$

$$\text{a } W \Subset \lambda^\cap \text{ with } a \in W \subset Y.$$

We wish to prove that $\omega' = \omega$.

λ is a subcollection of ω. Then, so are $\lambda^\cap$ and ω'. (Use the first description of ω'.) We now have $\omega' \sqsubset \omega$.

If φ is any topology on X and $\lambda \sqsubset \varphi$, then $\omega \sqsubset \varphi$. Thus, we will have $\omega \sqsubset \omega'$ if only ω' is a topology. To prove that it is, we apply the second description of ω'. From it, we see immediately that ω' satisfies the requirements T1 and T3 for a topology. Now, let $Y_1, Y_2 \Subset \omega'$; we prove $Y_1 \cap Y_2 \Subset \omega'$. Take $a \in Y_1 \cap Y_2$. As $Y_1, Y_2 \Subset \omega'$, there exist $W_1, W_2 \Subset \lambda^\cap$ with $a \in W_1 \subset Y_1$ and $a \in W_2 \subset Y_2$. The nice thing about $\lambda^\cap$ is that from $W_1 \Subset \lambda^\cap$ and $W_2 \Subset \lambda^\cap$ it follows that $W_1 \cap W_2 \Subset \lambda^\cap$. Further, we obviously have $a \in W_1 \cap W_2 \subset Y_1 \cap Y_2$. This concludes our proof. ∎

In each of the examples in Example 12.2, we have determined a topology ω by giving an explicit condition that is necessary and sufficient for a set to belong to ω. It is comparatively rare that one can do so: mostly, one can only describe a subbase.

Actually, for many purposes a subbase contains enough information and knowledge of the complete topology is redundant. A case in point is convergence, as we will see in the next theorem. (Another instance is continuity; Theorem 12.18.)

Theorem 12.14
Let λ be a subbase for a topology ω on X. Let $(x_\tau)_{\tau \in T}$ be a net in X and $x \in X$. Suppose that for every $W \in \lambda$ with $x \in W$, we have

$$x_\tau \in W \quad \text{for large } \tau.$$

Then the net converges to x.

Proof
Take $U \in \omega$ with $x \in U$. With $\lambda^\cap$ as in Theorem 12.13 we know that there is a $W \in \lambda^\cap$ for which $x \in W \subset U$. If $W = X$, then, trivially, $x_\tau \in U$ for all τ. Otherwise, W is an intersection $W_1 \cap \cdots \cap W_N$ of sets belonging to λ. For every W_n, there is a $\tau_n \in T$ such that

$$\tau \succ \tau_n \implies x_\tau \in W_n.$$

Take $\tau_0 \in T$ with $\tau_0 \succ \tau_n$ $(n = 1, \ldots, N)$. Then,

$$\tau \succ \tau_0 \implies x_\tau \in W_1 \cap \cdots \cap W_N = W \subset U. \qquad \blacksquare$$

Example 12.15
[Reprise and extension of Example 11.2(v).] Let S be a set and let F be a set of realvalued functions on S (e.g., S is $[0, 1]$ and F consists of all bounded functions $[0, 1] \to \mathbb{R}$). A net $(f_\tau)_{\tau \in T}$ in F is said to *converge pointwise* to a function $f \in F$ if

$$f_\tau(s) \to f(s) \quad \text{for every } s \in S.$$

We show that there is a topology ω on F such that ω-convergence is precisely pointwise convergence. (Of course, there is, at most, *one* such topology.)

For every open set $U \subset \mathbb{R}$ and every $s \in S$, let

$$U_s := \{f \in F : f(s) \in U\}.$$

The sets U_s form a collection λ of subsets of F:

$$\lambda := \{U_s : U \subset \mathbb{R} \text{ open}, s \in S\}.$$

This λ is a subbase for some topology ω on F.

Take a net $(f_\tau)_{\tau \in T}$ in F and let $f \in F$.

Suppose $f_\tau \overset{\omega}{\longrightarrow} f$. Let $s \in S$. If U is an open subset of $\mathbb{R}$ that contains $f(s)$, then $f \in U_s \in \lambda \sqsubset \omega$; hence, for large τ we have $f_\tau \in U_s$, i.e., $f_\tau(s) \in U$. Consequently, $f_\tau \to f$ pointwise.

Conversely, assume $f_\tau \to f$ pointwise; we prove ω-convergence. Let $f \in U_s \in \lambda$; by the previous theorem, we only have to show that $f_\tau \in U_s$

for large λ. That is not difficult to do: U is an open subset of $\mathbb{R}$ containing s, hence $f_\tau(s) \in U$ for large τ.

Thus, ω has the desired property. We call ω the *topology of pointwise convergence.*

Next, we look at continuity:

Theorem 12.16
Let (X,ω) and (X',ω') be topological spaces; let $f:X \to X'$. Then the following conditions on f are equivalent:

(α) *If $x_\tau \to x$ in X, then $f(x_\tau) \to f(x)$ in X'.*
(β) *If $V \in \omega'$, then $f^{-1}(V) \in \omega$.*
(γ) *If $A \subset X'$ is closed, then $f^{-1}(A)$ is closed in X.*

Proof
(α) $\Longrightarrow$ (γ). Let $x \in X$ be adherent to $f^{-1}(A)$; we are done if we can prove that $x \in f^{-1}(A)$ (last part of 12.7). There is a net $(x_\tau)_{\tau \in T}$ in $f^{-1}(A)$ that converges to x; then, $f(x_\tau) \to f(x)$. As $f(x_\tau) \in A$ for all τ and A is closed, we have $f(x) \in A$, i.e., $x \in f^{-1}(A)$.
(γ) $\Longrightarrow$ (β) is an elementary exercise.
(β) $\Longrightarrow$ (α) Let $x_\tau \to x$ in X. Let $V \in \omega'$ and $f(x) \in V$. Then x lies in the open subset $f^{-1}(V)$ of X. Hence, for large $\tau \in T$, we have $x_\tau \in f^{-1}(V)$, i.e., $f(x_\tau) \in V$. ∎

12.17
If (X, ω) and (X', ω') are topological spaces and $f : X \to X'$, we call f *continuous* (or, in times of stress, *ω-ω'-continuous*) if it satisfies any one, hence all, of the conditions above. [The textbooks usually choose condition (β).]

Observe that for $b \in X'$, the constant map

$$x \longmapsto b \qquad (x \in X)$$

is always continuous. If ω is the discrete topology [Example 12.2(ii)], then every map $X \to X'$ is continuous.

If (X, ω), (X', ω') , and $(X,'' \omega'')$ are topological spaces, a composition of a continuous map $X \to X'$ and a continuous map $X' \to X''$ is continuous.

Subbases can be a help in determining continuity:

Lemma 12.18
Let (X,ω) and (X',ω') be topological spaces; let $f:X \to X'$. Suppose there is a subbase λ' for ω' such that

$$W \in \lambda' \implies f^{-1}(W) \in \omega.$$

Then, f is continuous.

Proof
Let φ' be the collection of all subsets W of X' for which $f^{-1}(W) \in \omega$. It is a simple set-theoretic exercise to show that φ' is a topology. Furthermore, $\lambda' \sqsubset \varphi'$. Hence, $\omega' \sqsubset \varphi'$, i.e.,

$$W \in \omega' \implies f^{-1}(W) \in \omega.$$

This means that f satisfies (β) of Theorem 12.16.   ∎

12.19
A topology ω' in X is *weaker* than ω (and ω is *stronger* than ω') if $\omega' \sqsubset \omega$, i.e., if the identity map of X is ω-ω'-continuous. If ω' is weaker than ω and not equal to it, then ω' is *strictly weaker* than ω (and ω is *strictly stronger* than ω'). Thus, in $\mathbb{R}$, the Euclidean topology is strictly weaker than the discrete topology but strictly stronger than the indiscrete one.

If (X, ω) and (X', ω') are topological spaces, a map $f : X \to X'$ is called a *homeomorphism* if

$$\begin{bmatrix} f \text{ is bijective, and} \\ \text{for all } Y \subset X : \quad Y \in \omega \iff f(Y) \in \omega'. \end{bmatrix}$$

(See 4.17.) This is the case if and only if

$$\begin{bmatrix} f \text{ is bijective, and} \\ f \text{ and } f^{-1} \text{ are continuous.} \end{bmatrix}$$

Two topological spaces, (X, ω) and (X', ω'), are said to be *homeomorphic* if such a homeomorphism exists. For instance, under the Euclidean topologies, the intervals $[0, 1]$ and $[0, 2]$ are homeomorphic. If (X, d) and (X', d') are metric spaces, then a *surjective* isometry $X \to X'$ is a homeomorphism relative to the topologies determined by the metrics d and d'.

If (X, d) and (X', d') are metric spaces and if we have a bijective $f : X \to X'$ that is an isometry, then the inverse map $f^{-1} : X' \to X$ automatically is an isometry too. One might think for a moment that, similarly, the inverse of a bijective continuous map must be continuous. Even for metric spaces, however, that is false: Consider the case $X = X' = \mathbb{R}$ where f is the identity map, d is the trivial metric $[d(x, y) = 1$ as soon as $x \neq y]$, and d' is the Euclidean metric. (See, however, Corollary 13.23.)

12.20
If (X, d) is a metric space, then every subset Y of X is naturally endowed with a metric d_Y:

$$d_Y(x, y) := d(x, y) \qquad (x, y \in Y).$$

[See 5.4(v).] A similar thing can be done for topological spaces.

Let ω be a topology on a set X and let $Y \subset X$. Consider the collection of sets

$$\omega_Y := \{U \cap Y : U \in \omega\}.$$

It is not difficult to see that ω_Y is a topology on Y. (For example, $Y \in \omega_Y$ because $Y = X \cap Y$ and $X \in \omega$.) This topology is known as the *relative topology*, the *restriction topology*, the *subspace topology*, or the *induced topology*.

Henceforth, subsets of topological spaces will tacitly be endowed with the relative topology.

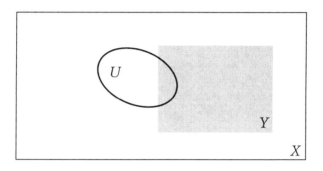

Theorem 12.21
Let (X,ω) be a topological space, $Y \subset X$.

(i) *If $(x_\tau)_{\tau \in T}$ is a net in Y and $x \in Y$, then*

$$x_\tau \xrightarrow{\ \omega_Y\ } x \quad \Longleftrightarrow \quad x_\tau \xrightarrow{\ \omega\ } x.$$

(ii) *Let (X',ω') be a topological space; let $f:X \to X'$ be ω-ω'-continuous. Then the restriction of f to Y is ω_Y-ω'-continuous.*

Proof

(i) Suppose we have ω-convergence. Take $V \in \omega_Y$ and $x \in V$. There is a $U \in \omega$ with $V = U \cap Y$. For large τ, we have $x_\tau \in U$, and, of course, $x_\tau \in Y$; hence, $x_\tau \in V$.

Conversely, if we start out with ω_Y-convergence and if $x \in U \in \omega$, then $x \in U \cap Y \in \omega_Y$, so that $x_\tau \in U \cap Y \subset U$ for large τ.

(ii) follows directly from (i).   ∎

Corollary 12.22
Let (X, d) be a metric space and $Y \subset X$. If ω is the d-topology on X, then ω_Y is the d_Y-topology on Y.

Proof

If $(x_\tau)_{\tau \in T}$ is a net in Y and $x \in Y$, then

$$x_\tau \xrightarrow{\omega_Y} x \quad \Longleftrightarrow \quad x_\tau \xrightarrow{\omega} x$$

$$\Longleftrightarrow \quad x_\tau \xrightarrow{d} x \quad \Longleftrightarrow \quad x_\tau \xrightarrow{d_Y} x. \qquad \blacksquare$$

A special case: If $Y \subset X \subset \mathbb{R}^N$, then the Euclidean topology on Y is the restriction of the one on X.

Theorem 12.23 (Glue Lemma; first version)

Let (X,ω) and (X',ω') be topological spaces; let $f:X \to X'$. Suppose there exists a finite sequence $X_1, \ldots, X_N$ of closed subsets of X with

$$\begin{bmatrix} X = X_1 \cup \ldots \cup X_N; \\ \text{for each } n \text{ the restriction of } f \text{ to } X_n \text{ is continuous.} \end{bmatrix}$$

Then f is continuous.

Proof

Let $A \subset X'$ be ω'-closed; we prove $f^{-1}(A)$ to be ω-closed. To do this, it certainly suffices to show that each intersection $f^{-1}(A) \cap X_n$ is ω-closed.

Take $n \in \{1, \ldots, N\}$. If $(x_\tau)_{\tau \in T}$ is a net in $f^{-1}(A) \cap X_n$, ω-converging to $x \in X$, then $x \in X_n$ because X_n is ω-closed and $f(x_\tau) \to f(x)$ because the restriction of f to X_n is continuous. As $f(x_\tau) \in A$ for all τ, we have $f(x) \in A$ and $x \in f^{-1}(A) \cap X_n$. $\qquad \blacksquare$

See Exercise 12.G for another form of the Glue Lemma.

Extra: Map Coloring

One of the famous problems of mathematics was, in a sense, settled in 1976.

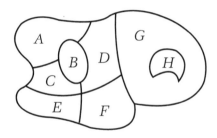

Above, we have a map of a continent on some distant planet. There are eight countries on it. Suppose you are asked to color this map in the usual fashion, so that no two adjacent countries get the same color. If you have eight crayons at your disposal you can do it, of course. You don't really need eight: There is no reason why *A*, *E* and *H* should not get the same color, so six crayons would be enough. Can you do it with fewer colors? Yes: Let *A*, *E* and *H* be red, *B* and *F* blue, *C* and *G* yellow, and *D* purple. Four colors suffice. You cannot do it with three, since *A*, *B*, *C* and *D* all border each other.

In 1852, Francis Guthrie, a student at Edinburgh, noticed that for all maps he tried four colors were enough. The "Four Color Problem" was born: Are four colors sufficient for every possible map? (It is understood that all countries consist of one piece. Two countries must get distinct colors if they have a common frontier; in the map above, the countries *C* and *F* may be colored the same.)

Guthrie's brother communicated the problem to his mathematics teacher, Augustus de Morgan, and so it reached a wider audience. It was soon proved by Percival Heawood that no map requires more than five colors. After that, many partial results and false proofs were published, but only in 1976, Kenneth Appel and Wolfgang Haken, two mathematicians then at the University of Illinois, showed that, indeed, four colors always suffice. Since then, we have the Four Color Theorem.

Many extensions of the theorem have been proved. For instance, if every country consists of *two* pieces (that have to get the same color, of course), then twelve colors are enough, and there are maps that need twelve. See Ian Stewart's article *The Rise and Fall of the Lunar M-pire* in the April 1993 issue of the *Scientific American*.

We have formulated the problem for a continent, but planets are balls. How many colors does one need for a globe? It is not difficult to see that the answer is the same: four is enough, three is not.

You can also consider the problem on other surfaces, such as the torus (the surface of a donut):

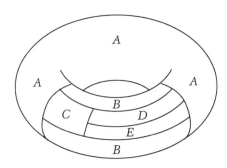

As you see, here you have five countries, each bordering each: four colors are not sufficient. Indeed, for a torus the magical number is not 4 but 7! This "Seven Color Theorem" for a torus was proved by Haerwood as early as the end of the Nineteenth century.

It has been shown by Ringel (1959) that only six colors are needed for the Klein bottle. (See 14.7.)

All of the above is part of "Combinatorial Topology." If you are interested, read more about it in *Graphs and Digraphs* by G. Chartrand and L. Lesniak, or in the delightful *Initiation to Combinatorial Topology* by M. Fréchet and K. Fan.

We said that the Four Color Problem was settled "in a sense." The trouble with the solution is this. Using conventional (but far from simple) mathematical methods, Appel and Haken reduced the problem to a straightforward mathematical calculation that still had to be performed. However, that calculation is so complicated that no one in his (or her) lifetime could possibly do it. A machine could, and did.

Does that count as a proof? Why not? you may say. Mathematicians use tools such as pencils and chalk all the time; then why not a computer? But our pencils do not tell us that something is true; we are not asked to believe them. Is a proof a proof if it requires trusting a computer?

Further Reading

Chartrand, G. and L. Lesniak, *Graphs and Digraphs*, Wadsworth & Brooks/Cole, Pacific Grove, California: 1986.

Fréchet, M. and K. Fan, *Initiation to Combinatorial Topology*, Prindle, Weber and Schmidt, Pacific Grove, California: 1967.

Lam, C.W.H., How Reliable Is a Computer-Based Proof? *Mathematical Intelligencer* 12 (1990), 8–12.

Exercises

12.A. Which of the following collections of subsets of $[0, 1]$ are topologies on $[0, 1]$?
   (a) $\omega_1 := \{A : A \subset (0, 1) \text{ or } A = [0, 1]\}$.
   (b) $\omega_2 := \{A : A \subset \mathbb{Q} \cap [0, 1] \text{ or } \mathbb{Q} \cap [0, 1] \subset A\}$.
   (c) $\omega_3 := \{A : AA \subset A\}$. (Here $AA := \{xy : x \in A, y \in A\}$.)
   (d) $\omega_4 := \{A : (A{+}A) \cap [0, 1] \subset A\}$. (Here $A{+}A := \{x{+}y : x \in A, y \in A\}$.)
   (e) $\omega_5 := \{A : 0 \notin A \text{ or } A = [0, 1]\}$.
   (f) $\omega_6 := \{A : 0 \in A \text{ or } A = \varnothing\}$.

12.B. We consider two topologies on $\mathbb{R}$: the topology ω generated by the Euclidean metric and the topology ω' having the collection of all intervals $[a, b)$ as a (sub)base.

Show that a function $f : \mathbb{R} \to \mathbb{R}$ is ω'-ω-continuous if and only if

$$\lim_{x \downarrow a} f(x) = f(a) \quad \text{for every } a \in \mathbb{R}.$$

Show that all ω-ω'-continuous functions $\mathbb{R} \to \mathbb{R}$ are constant.

12.C. (i) Take K as in Example 9.11. In Exercise 9.G, we have seen that for sequences in K uniform convergence coincides with pointwise convergence. Show that the same is true for nets.

(ii) By

$$[0, 1]^{\mathbb{N}}$$

we indicate the set of all functions $\mathbb{N} \to [0, 1]$. To each $f \in [0, 1]^{\mathbb{N}}$ we assign $f' \in K$ by

$$f'(n) := n^{-1}f(n) \qquad (n \in \mathbb{N}).$$

For $f, g \in [0, 1]^{\mathbb{N}}$, put

$$d(f, g) := \|f'-g'\|_{\infty}.$$

Show that d is a metric on $[0, 1]^{\mathbb{N}}$ and that for nets in $[0, 1]^{\mathbb{N}}$ d-convergence is the same as pointwise convergence.

The topological space $[0, 1]^{\mathbb{N}}$, provided with the d-topology, is called the *Hilbert Cube*. (Its topology is characterized by the property that topological convergence is precisely pointwise convergence.)

12.D. Let (X, ω) be a topological space and Y a subset of X. We provide Y with the relative topology ω_Y.

(i) Prove that for a subset Z of Y, *the ω_Y-closure of Z is just the intersection of Y with the ω-closure of Z*. Give an example showing that the ω_Y-interior of Z may *not* be the intersection of Y with the ω-interior of Z.

(ii) Show: *If Y is ω-closed, then a subset of Y is ω_Y-closed if and only it it is ω-closed. If Y is ω-open, then a subset of Y is ω_Y-open if and only if it it is ω-open.*

12.E. (On the relative topology.) Let (X, ω) be a topological space, $Y \subset X$. Show that *if φ is a (sub)base for ω, then*

$$\varphi_Y := \{U \cap Y : U \in \varphi\}$$

is a (sub)base for ω_Y.

12.F. Let X be a topological space and $C(X)$ the set of all continuous functions $X \to \mathbb{R}$.

(i) Prove

$$\text{if } f, g, \in C(X), \text{ then } f+g \in C(X), fg \in C(X), \text{ and } f \vee g \in C(X).$$

($f+g$, fg, and $f \vee g$ are the functions $x \mapsto f(x)+g(x)$, $x \mapsto f(x)g(x)$, and $x \mapsto f(x) \vee g(x)$, respectively. Hint: Use nets.)

(ii) Prove: if $f_1, f_2, \ldots \in C(X)$ and if f is a function on X such that $f_n \to f$ uniformly (8.4 and Theorem 8.6), then $f \in C(X)$.

12.G. (i) Prove the **Glue Lemma, Second Version**: *Let f be a map of a topological space X into a topological space X' and let there exist a collection φ of open subsets of X whose union is X, such that for every $Y \in \varphi$ the restriction of f is a continuous map $Y \to X'$. Then, f is continuous.*

(ii) Show that the theorem obtained from the above by changing "open" into "closed" is false.

12.H. Let X be a topological space. Let f be a function $X \to \mathbb{R}$ such that for all $a \in \mathbb{R}$, the sets

$$\{x : f(x) > a\} \quad \text{and} \quad \{x : f(x) < a\}$$

are open. Prove that f is continuous.

12.I. (See Exercise 6.M.) Let X be a topological space. For $Y \subset Z \subset X$, we say that Y is *dense* in Z if $Z \subset Y^-$. Thus, Y is dense in X if and only if $Y^- = X$.

(i) For $Y \subset X$, prove that *the following are equivalent*:
  (α) *Y is dense in X.*
  (β) *Every element of X is the limit of a net in Y.*
  (γ) *Every nonempty open subset of X contains a point of Y.*

(ii) Let Y be a dense subset of X and let U be an open subset of X. Show that $Y \cap U$ is dense in U.

(iii) Show that if $Y \subset Z \subset X$, if Y is dense in Z and Z is dense in X, then Y is dense in X.

(iv) Show [directly or from (ii) and (iii)] that if U_1 and U_2 are open dense subsets of X, then $U_1 \cap U_2$ is (open and) dense.

12.J. Let X be a topological space.

(i) Show that if Y is a closed subset of X with empty interior, then $\partial Y = Y$.

(ii) Show that for every subset Y of X, ∂Y is closed and $\partial\partial Y$ has empty interior. [Hence, with (ii), $\partial\partial\partial Y = \partial\partial Y$.]

(iii) Give an example of a subset Y of $\mathbb{R}$ with $\partial\partial Y \neq \partial Y$ (Euclidean topology).

12.K. Let (X, ω) be a topological space with the property

$$\text{if } U_1, U_2, \ldots \in \omega, \text{ then } \bigcap_n U_n \in \omega.$$

(A nontrivial example is presented in Exercises 11.B and 13.G.)

Show that a function $f : X \to \mathbb{R}$ is continuous if and only if $\{x : f(x) = a\}$ is open for every $a \in \mathbb{R}$.

Show that the pointwise limit of any sequence of continuous functions on X is continuous.

13 CHAPTER | Compactness and the Hausdorff Property

Compact Spaces

13.1

Up to this point, it has been easy to transfer definitions and results from the theory of metric spaces to the theory of topological spaces. In fact, only once (in connection with Theorem 11.18) we have had a problem. With the introduction of compactness, we encounter a new difficulty. For metric spaces, we have defined sequential compactness, a concept going back to the Bolzano-Weierstrass Theorem (3.8). Its strength and applicability rest on Theorems 9.10 and 9.14. Theorem 9.10, however, is not salvageable in the setting of topological spaces. Indeed, the definition of "totally bounded" is contingent on the existence of a metric. As to Theorem 9.14, the definition of "sequentially compact" can be carried to the stage of topological spaces verbatim: A topological space X is called *sequentially compact* if every sequence of elements of X has a convergent subsequence. Similarly, the other players in Theorem 9.14 can act a part: Conditions (β) and (γ) of Theorem 9.14 make perfectly good sense in a topological space.

Unfortunately, if we set the stage like that, the play has no happy end. Although (β) and (γ) are mutually equivalent, we will obtain an example of a sequentially compact space without (γ) (Example 13.5) and much later, in Example 17.13 an example of a space that has property (γ) without being sequentially compact. So we have a dilemma. Should we focus on sequential compactness or on property (γ)?

This dilemma was more acute in the days when the theory of topological spaces was still in its infancy. Since then, (γ) has become the accepted norm, whereas sequential compactness has turned into a bit player.

13.2

A topological space X (or its topology) is called *compact* if every cover of X by open sets has a finite subcover. (See 9.8 for the terminology.)

Examples 13.3

(i) For metric spaces, compactness is equivalent to sequential compactness. (This is Theorem 9.14.) (However, see Examples 13.5 and 17.13.) The compactness of $[0, 1]$ is known as the *Heine-Borel Theorem*.

(ii) A space with the indiscrete topology [Example 12.2(iii)] is compact.

(iii) A space with the discrete topology is compact if and only if it is finite.

(iv) Under the topologies of (v) and (vi) of Example 12.2, $\mathbb{R}$ is not compact.

(v) For any set X, the topology introduced in Example 12.2(iv) is compact.

Theorem 13.4

For a topological space X, the following are equivalent:

(α) *X is compact.*

(β) *Every finitely bound collection (see 9.13) of closed subsets of X has nonempty intersection.*

Proof

We leave the proof to you. ∎

Example 13.5

A sequentially compact space that is not compact:

For a function $f : [0, 1] \to \mathbb{R}$, set $S(f) := \{t \in [0, 1] : f(t) \neq 0\}$.

Let X be the set of all functions $f : [0, 1] \to \mathbb{R}$ for which

$$\begin{cases} f(t) \in [0, 1] \text{ for every } t \in [0, 1]; \\ S(f) \text{ is countable.} \end{cases}$$

X is a subset of the set of *all* functions on $[0, 1]$. We endow X with the topology ω of pointwise convergence (see Example 12.15). Then, for $f, f_1, f_2, \ldots \in X$, we have

$$f_n \xrightarrow{\omega} f \iff f_n(t) \longrightarrow f(t) \text{ for all } t \in [0, 1].$$

To prove sequential compactness, take a sequence $f_1, f_2, \cdots$ in X. The set $S := S(f_1) \cup S(f_2) \cup \cdots$ is countable. By an easy extension of Corollary

9.12, the sequence of restrictions $f_1|_S, f_2|_S, \ldots$ has a subsequence that converges pointwise (on S). It follows that the sequence $f_1, f_2, \ldots$ itself has a subsequence $g_1, g_2, \ldots$ such that $\lim_{n\to\infty} g_n(t)$ exists for all $t \in S$. Since $g_n(t) = 0$ $(n \in \mathbb{N})$ as soon as $t \notin S$, we see that the sequence $g_1, g_2, \ldots$ converges pointwise on all of $[0, 1]$, hence converges in the sense of ω.

To see that X is not compact, with every $t \in [0, 1]$ we associate the set

$$A_t := \{f \in X : f(t) = 1\}.$$

The sets A_t $(t \in [0, 1])$ form a collection of closed subsets of X that is finitely bound but has empty intersection.

13.6
Thus, the conditions mentioned in Theorem 13.4 are not equivalent to

> every sequence in X has a convergent subsequence.

That was to be expected; sequences will have to be replaced by nets. Indeed, there is a notion of "subnet" that makes compactness equivalent to

> every net in X has a convergent subnet.

Such a notion turns out be quite sophisticated. Let us see how we should define subnets. (If you are not particularly interested in the motivation of the definition, you may as well skip the discussion and continue reading 13.9.)

First, subnets ought to have the nice properties of subsequences. It seems reasonable to ask that for any net $(x_\tau)_{\tau\in T}$, we have the following:

(a) $(x_\tau)_{\tau\in T}$ is a subnet of itself.
(b) Every subnet of a subnet of $(x_\tau)_{\tau\in T}$ is itself a subnet of $(x_\tau)_{\tau\in T}$.
(c) If $(x_\tau)_{\tau\in T}$ converges to a point x, then so does every subnet.
   Moreover, subnets should furnish a description of compactness:
(d) A topological space is compact if and only if every net has a convergent subnet.

13.7
Our first attempt at a definition: A subnet of $(x_\tau)_{\tau\in T}$ is a net $(x_\tau)_{\tau\in S}$, where S is a subset of T. On second thought, we add the condition that S be directed. Even so, however, this definition fails to ensure (c), even for $T = \mathbb{N}$: The net $(n^{-1})_{n\in\mathbb{N}}$ in $\mathbb{R}$ converges to 0, the net $(n^{-1})_{n\in\{1,2,3\}}$ does not.

Looking back at the definition of "subsequence" (3.1), we see that it might be wiser to define a subnet of $(x_\tau)_{\tau\in T}$ to be a net $(x_{\alpha(\tau)})_{\tau\in T}$, where α is a suitable map $T \to T$. For subsequences, we required α to be strictly increasing. For subnets, that will not do. To see this, let T be the interval $(0, 1)$, directed by $\leq$, and $\alpha(\tau) = \tau/2$ $(\tau \in T)$; then, $(e^\tau)_{\tau\in T}$ is a net in $\mathbb{R}$ that converges to e, but $(e^{\alpha(\tau)})_{\tau\in T}$ converges to $e^{\frac{1}{2}}$.

If a subnet of $(x_\tau)_{\tau \in T}$ is going to be $(x_{\alpha(\tau)})_{\tau \in T}$ with $\alpha : T \to T$, what property of α will guarantee (c)? From the existence of a $\tau_0 \in T$ satisfying a formula of the type

$$\tau \succ \tau_0 \quad \Longrightarrow \quad x_\tau \in \dots,$$

we wish to be able to infer that there is a τ_1 with

$$\tau \succ \tau_1 \quad \Longrightarrow \quad x_{\alpha(\tau)} \in \dots.$$

This will be okay if

for every τ_0, there exists a τ_1 $\qquad (*)$

such that $\tau \succ \tau_1 \quad \Longrightarrow \quad \alpha(\tau) \succ \tau_0$.

In the case of (sub)sequences ($T = \mathbb{N}$), this can be formulated as

$$\lim_{\tau \to \infty} \alpha(\tau) = \infty. \qquad (**)$$

For subsequences (see 3.1), we required α to be strictly increasing. That condition implies $(**)$, of course, but it is unnecessarily restrictive. What makes subsequences work is really $(**)$.

Accordingly, for subnets we are going to adopt $(*)$. Unfortunately, if we define the subnets of $(x_\tau)_{\tau \in T}$ to be the nets $(x_{\alpha(\tau)})_{\tau \in T}$ with $\alpha : T \to T$ satisfying $(*)$, we do not get our wish (d), as the following example shows.

Example 13.8
In Exercise 12.C, we have seen that there is a metric d on the set

$$X := [0, 1]^{\mathbb{N}}$$

such that d-convergence is pointwise convergence and that X is sequentially compact, hence compact. We define a net $(f_\tau)_{\tau \in T}$ in X as follows. For the index set T, we take the set of all nonempty countable subsets of $[0, 1]$, $\succ$ being $\supset$. For every $\tau \in T$, choose an $f_\tau : \mathbb{N} \to [0, 1]$ with

$$\tau = \{f(1), f(2), \dots\}.$$

Our object is to show that there is no $\alpha : T \to T$ satisfying $(*)$ of 13.7 and such that $(f_{\alpha(\tau)})_{\tau \in T}$ converges.

Suppose we have such an α and $f_{\alpha(\tau)} \to g \in X$. Choose a metric d for the topology of X. For every $n \in \mathbb{N}$, there is a $\sigma_n \in T$ such that

$$\tau \in T, \tau \supset \sigma_n \quad \Longrightarrow \quad d(f_{\alpha(\tau)}, g) < n^{-1}.$$

Now, $\sigma_\infty := \bigcup_n \sigma_n$ is an element of T. If $\tau \in T$ and $\tau \supset \sigma_\infty$, then $\tau \supset \sigma_n$ for all n, from which $f_{\alpha(\tau)} = g$. Thus,

$$\tau \in T, \tau \supset \sigma_\infty \quad \Longrightarrow \quad \alpha(\tau) = \{g(1), g(2), \dots\}. \qquad (1)$$

$[0, 1]$ being uncountable, there is an x in $[0, 1]$ that does not lie in $\{g(1), g(2), \dots\}$. Property $(*)$ of α implies the existence of a $\tau_1 \in T$ with

$$\tau \in T, \tau \supset \tau_1 \quad \Longrightarrow \quad \alpha(\tau) \supset \{x, g(1), g(2), \dots\}. \qquad (2)$$

But (2) *contradicts* (1).

The solution to our problem is drastic. We abandon the idea that the index set of the subnet has anything to do with that of the net. The following definition of "subnet" is not the only one that exists in the literature, but it will suffice for our purposes.

13.9

A net $(y_\sigma)_{\sigma \in S}$ is said to be a *subnet* of a net $(x_\tau)_{\tau \in T}$ if there exists a map $\alpha : S \to T$ such that

(a) $y_\sigma = x_{\alpha(\sigma)}$ for every $\sigma \in S$;
(b) for every $\tau_0 \in T$, there is a $\sigma_0 \in S$ with

$$\sigma \succ \sigma_0 \implies \alpha(\sigma) \succ \tau_0.$$

13.10

(i) We have a similar diagram as in 3.2:

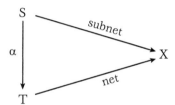

(ii) Property (b) of α may be interpreted as

$$\sigma \to \infty \implies \alpha(\sigma) \to \infty.$$

(iii) Let $(x_n)_{n \in \mathbb{N}}$ be a sequence. Every subsequence is a subnet, but there are also subnets that are not sequences, or even have uncountable index sets. For an example, we can take $S := [1, \infty)$ and define $\alpha(\sigma)$ to be the entire part of σ.

(iv) In Definition 13.9, we have used the same symbol $\succ$ for the direction in T as well as the direction in S. Such ambiguity generally is undesirable, but we tried to avoid cluttering up an already convoluted definition.

(v) Our definition fulfills our wishes (a), (b), and (c) of 13.6. In Theorem 13.13, we prove that it also yields (d). The following terminology will help.

13.11

Let $(x_\tau)_{\tau \in T}$ be a net in a topological space X. An element x of X is called a *cluster point* of the net if

for every open set U containing x and every $\tau_0 \in T$,

there is a $\tau \in T$ with $\tau \succ \tau_0$ and $x_\tau \in U$, which is the case if and only if

$$x \in \mathrm{clo}\{x_\tau : \tau \succ \tau_0\} \quad \text{for all } \tau_0 \in T.$$

Apparently, the cluster points of a net form a closed set. Every limit of a net is a cluster point.

Lemma 13.12
Let $(x_\tau)_{\tau \in T}$ be a net in a topological space X; let $x \in X$. Then x is a cluster point of the net if and only if the net has a subnet converging to x.

Proof
The "if" is virtually trivial. To prove the "only if," assume x is a cluster point of the net.

Let S be the set of all pairs (U, τ) where U is an open set containing x and $\tau \in T$, $x_\tau \in U$. The definition

$$(U', \tau') \succ (U, \tau) \quad \Longleftrightarrow \quad U' \subset U \text{ and } \tau' \succ \tau$$

renders S a directed set. (The inductivity of $\succ$ follows from the fact that x is a cluster point.) Define a net $(y_{(U,\tau)})_{(U,\tau) \in S}$ by

$$y_{(U,\tau)} := x_\tau.$$

This net is a subnet of the given net $(x_\tau)_{\tau \in T}$ via the map $\alpha : (U, \tau) \mapsto \tau$ and is easily seen to converge to x.   ∎

Theorem 13.13
A topological space X is compact if and only if every net in X has a cluster point (hence, if and only if every net has a convergent subnet.)

Proof

(I) Suppose X is compact; let $(x_\tau)_{\tau \in T}$ be a net. For $\tau \in T$, put

$$A_\tau := \mathrm{clo}\{x_{\tau'} : \tau' \succ \tau\}.$$

These sets A_τ form a finitely bound collection of closed sets. By compactness, its intersection is nonempty. But the intersection consists precisely of the cluster points of the net $(x_\tau)_{\tau \in T}$.

(II) Suppose every net has a cluster point. Let α be a finitely bound collection of closed sets. Let $\alpha^\cap$ be the collection of all finite intersections $A_1 \cap \cdots \cap A_N$ of sets belonging to α. With

$$A \succ B \quad \Longleftrightarrow \quad A \subset B \quad (A, B \in \alpha^\cap),$$

$\alpha^\cap$ is a directed set. For every $A \in \alpha^\cap$, choose an element x_A of A. Then, $(x_A)_{A \in \alpha^\cap}$ is a net, having a cluster point, x, say. For all $A \in \alpha^\cap$, we have $x \in \mathrm{clo}\{x_B : B \in \alpha^\cap, B \succ A\} \subset \mathrm{clo} A = A$. Hence, $x \in \bigcap \alpha$.   ∎

We have defined compactness as a property of topological spaces. There is also a notion of compactness of sets as subsets of a topological space.

13.14

Let (X, ω) be a topological space and let A be a subset of X. We call A ω-compact if it has the following property:

> for every subcollection φ of ω for which $A \subset \bigcup \varphi$,
>
> there exist $N \in \mathbb{N}$ and $U_1, \ldots, U_N \in \varphi$ with $A \subset U_1 \cup \ldots \cup U_N$.

For $A = X$, this definition is in accordance with the one given in 13.2.

Long ago (6.17), we observed that some topological notions are "relative": the set $[0, 1]$ is open if considered as a subset of the space $[0, 1] \cup [2, 3]$ but not as a subset of $\mathbb{R}$. Compactness is different:

Theorem 13.15
Let (X, ω) be a topological space and let $A \subset X$. Then A is ω-compact if and only if the topological space (A, ω_A) is compact.

Proof
Suppose A is ω-compact. Let φ be a cover of A with ω_A-open sets. We show that φ has a finite subcover.

Let

$$\varphi_1 := \{U \in \omega : U \cap A \in \varphi\}.$$

By the definition of ω_A, every set V belonging to φ is of the form $U \cap A$ for some $U \in \omega$; then, $V \subset U$ and $U \in \varphi_1$. Hence, $A \subset \bigcup \varphi_1$. Because A is ω-compact, there exist $U_1, \ldots, U_N \in \varphi_1$ such that

$$A \subset U_1 \cup \cdots \cup U_N.$$

But then,

$$A = (U_1 \cap A) \cup \cdots \cup (U_N \cap A)$$

and each $U_n \cap A$ belongs to φ. We have the desired finite subcover.

The proof of the converse (ω_A-compactness implies ω-compactness) is simpler; we leave it to you. ∎

We now collect some properties that have made compactness useful.

Theorem 13.16
Let (X, ω) be a compact topological space.

(i) *Every closed subset of X is compact. (Compare Theorem 9.2.)*

(ii) *If f is a continuous map of X into a topological space X' and A is a compact subset of X, then f(A) is compact. (Compare Exercise 9.F.)*

Proof

(i) Let $A \subset X$ be closed. Take a net $(x_\tau)_{\tau \in T}$ in A. By the compactness of X, the net has a subnet $(y_\sigma)_{\sigma \in S}$ that ω-converges to some $y \in X$. As A is closed, we have $y \in A$. Then, $y_\sigma \to y$ in the sense of ω_A. It follows that A is ω_A-compact, hence ω-compact.

(ii) Let ω' be the topology of X'; we prove $f(A)$ to be ω'-compact. Take a subcollection φ of ω' with $f(A) \subset \bigcup \varphi$. Then,

$$\{f^{-1}(U) : U \in \varphi\}$$

is a subcollection of ω whose union contains A. As A is ω-compact, there exist $U_1, \ldots, U_N \in \varphi$ with

$$A \subset f^{-1}(U_1) \cup \cdots \cup f^{-1}(U_N).$$

Then, $f(A) \subset U_1 \cup \ldots \cup U_N$. ∎

Corollary 13.17

If X is a nonempty compact space, then every continuous function $X \to \mathbb{R}$ is bounded and assumes a largest value. (Compare Theorem 9.3.)

Proof

Let $f : X \to \mathbb{R}$ be continuous. By Theorem 13.16(ii), $f(X)$ is a compact subset of $\mathbb{R}$. Then, $f(X)$ is bounded and closed, and therefore contains its supremum. ∎

Hausdorff Spaces

Our list of useful properties of compact spaces is not yet complete. There will be more in Theorem 13.22 and Corollary 13.23, after the introduction of the Hausdorff property.

This property is so natural that Felix Hausdorff made it a part of *his* definition of a topology. We have seen that a net in a topological space may have more than one limit. In spaces with the Hausdorff property, this cannot happen.

13.18

A topological space X is said to have the *Hausdorff property* (or to be a *Hausdorff space*) if for every pair of elements a, b of X with $a \neq b$, there exist open sets U and V with

$$\left[\begin{array}{l} a \in U \text{ and } b \in V, \\ U \cap V = \varnothing. \end{array}\right.$$

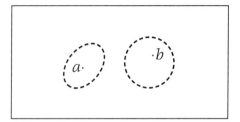

Example 13.19

(i) Every metrizable space is Hausdorff. Indeed, let d be a metric on X. Take $a, b \in X$, $a \neq b$. Let

$$r := \tfrac{1}{2} d(a, b).$$

Then $r > 0$, and the open balls $B_r(a)$ and $B_r(b)$ have empty intersection.

(ii) Let S be any set and $F(S)$ the space of all functions on S. Under the topology of pointwise convergence (Example 12.15), $F(S)$ is Hausdorff. Proof. Take $f, g \in F(S)$ and $f \neq g$. There is an $s_0 \in S$ with $f(s_0) \neq g(s_0)$. There exist open intervals $I, J \subset \mathbb{R}$ with $f(s_0) \in I$, $g(s_0) \in J$, and $I \cap J = \varnothing$. The sets $U := \{h \in F(S) : h(s_0) \in I\}$ and $V := \{h \in F(S) : h(s_0) \in J\}$ are open in $F(S)$. (They belong to the subbase described in Example 12.15.) Clearly, $f \in U$, $g \in V$, and $U \cap V = \varnothing$.

(iii) The space described in Example 12.2(v) is *not* Hausdorff, the one in (vi) is.

Theorem 13.20
A topological space X is Hausdorff if and only if no net in X has more than one limit.

Proof
First, let X be Hausdorff. *Suppose* we have a net $(X_\tau)_{\tau \in T}$ in X with limits a and b while $a \neq b$. Choose open $U, V \subset X$ such that $a \in U$, $b \in V$, and $U \cap V = \varnothing$. There exist $\tau_1, \tau_2 \in T$ for which

$$\tau \succ \tau_1 \implies x_\tau \in U,$$

$$\tau \succ \tau_2 \implies x_\tau \in V.$$

But there is a $\tau \in T$ with $\tau \succ \tau_1$ and $\tau \succ \tau_2$. Then $x_\tau \in U \cap V$. *Contradiction.*

Now assume that X does not have the Hausdorff property. Let $a, b \in X$ be such that $a \neq b$ but $U \cap V \neq \varnothing$ for all open sets U and V with $a \in U$ and $b \in V$. Consider the set

$$T := \{(U, V) : U, V \text{ are open}, a \in U, b \in V\}.$$

Define a direction $\succ$ on T by

$$(U', V') \succ (U, V) \iff U' \subset U \text{ and } V' \subset V.$$

For every $(U, V) \in T$, choose an element $x_{(U,V)}$ of $U \cap V$. Then, we have a net $(x_{(U,V)})_{(U,V) \in T}$. If U_0 is any open set containing a, then $(U_0, X) \in T$ and

$$(U, V) \succ (U_0, X) \implies x_{(U,V)} \in U \subset U_0.$$

Hence, our net converges to a. Similarly, it converges to b. Thus, we have a net with at least two limits. ∎

Theorem 13.21
In a Hausdorff space, every finite set is closed.

Proof
Let X be a Hausdorff space. It is enough to show that for every $a \in X$ the set $\{a\}$ is closed, i.e., $X \backslash \{a\}$ is open. But $X \backslash \{a\}$ obviously is a union of open sets. ∎

The combination of compactness and Hausdorffness works wonders:

Theorem 13.22
(i) *A compact subset of a Hausdorff space is closed.*
(ii) *A continuous map from a compact space into a Hausdorff space maps closed sets onto closed sets.*

Proof
(i) Let A be a compact set in a Hausdorff space (X, ω). Let $(x_\tau)_{\tau \in T}$ be a net in A, ω-converging to $x \in X$; we are done if we can prove $x \in A$. As (A, ω_A) is compact, the net has a subnet that ω_A-converges to some $y \in A$. This subnet is ω-convergent to x but also to y [Theorem 12.21(i)]. Limits being unique in X, we obtain $x = y \in A$.
(ii) Let X be a compact space and X' a Hausdorff space. Let $f : X \to X'$ be continuous and let A be a closed subset of X. Then A is compact [Theorem 13.16(i)]; hence, $f(A)$ is compact [Theorem 13.16(ii)]; hence, $f(A)$ is closed in X' [Theorem 13.22(i)]. ∎

Corollary 13.23
A continuous bijective map from a compact space onto a Hausdorff space is a homeomorphism.

Proof

Let f be a continuous bijection from a compact space X onto a Hausdorff space X', and let $g : X' \to X$ be its inverse. We wish to prove the continuity of g. Take a closed subset A of X; we are done if $g^{-1}(A)$ is closed in X'. But $g^{-1}(A) = f(A)$ and $f(A)$ is closed [Theorem 13.22(ii)]. ∎

Two other important results on compact Hausdorff spaces are Urysohn's Metrization Theorem (15.16) and a version of the Baire Category Theorem (15.17).

Extra: Hausdorff and the Measure Problem

Felix Hausdorff (1868-1942) was born in Wroclaw, now in Poland, but in those days was in the Kingdom of Prussia. He started his career as an astronomer, working at the observatory of Leipzig, Germany, and writing a thesis on the absorption of light in the atmosphere. Starting in 1901, he was a professor of mathematics at the universities of Leipzig and, after that, Bonn. He always remained interested in applied as well as pure mathematics. (He also wrote poetry and philosophy, under the pseudonym of Paul Mongré.) In 1942, faced by the threat of deportation to one of Hitler's concentration camps, he committed suicide.

To us, Hausdorff is of importance as one of the Founding Fathers of Topology, as pure a branch of mathematics as you can get.

His *Mengenlehre* ("Set Theory") of 1914 marks one of those instances where a whole subject comes of age by the appearance of a single book. Before that, the knowledge of Topology, or Analysis Situs, as it was called, had been scattered and diffuse. Hausdorff was able to pick the essentials of it, make the right generalizations, add a substantial amount of new material, and set the standard for times to come. Even now, his book is still a readable introduction to Topology.

We have already mentioned two of this contributions: the Hausdorff metric (Exercise 8.D) and, of course, the Hausdorff property. Hausdorff also introduced a notion of "dimension" to metric spaces. By his definition, the dimension of $\mathbb{R}^N$ is N, as is only reasonable; $\mathbb{Q}$ and $\mathbb{R}\backslash\mathbb{Q}$ have dimensions 0 and 1, respectively. Further, he studied what is now called the Hausdorff Maximal Principle, a set-theoretic statement equivalent to the Axiom of Choice and Zorn's Lemma; see the Extra of Chapter 17.

We cannot refrain from mentioning a bit of work Hausdorff did in a different field of mathematics, Measure Theory.

Intuitively speaking, every bounded subset X of $\mathbb{R}^2$ has a well-determined area, $\mu(X)$, say, and for all X and Y:

(a) $\mu(X) \geq 0$,
(b) $\mu(X \cup Y) = \mu(X) + \mu(Y)$ if $X \cap Y = \varnothing$,
(c) $\mu(T(X)) = \mu(X)$ if T is an isometry $\mathbb{R}^2 \to \mathbb{R}^2$,
(d) $\mu([0, 1] \times [0, 1]) = 1$.

In the same way, one expects every bounded subset of $\mathbb{R}^3$ to have a volume, obeying similar rules. (The word "measure" in the title is a general term covering "area," "volume," and similar notions.) It is quite another matter, however, to *prove* that such an "area function" $X \mapsto \mu(X)$ and a "volume function" exist. Hausdorff obtained the following results:

(I) There exists an area function satisfying (a)-(d). That may sound uninteresting, but there is a snag: There exist many area functions. According to all of them, the area of a rectangle is the product of the sides and the area of a disk with radius R is πR^2, but you can make exotic sets to which distinct area functions assigns distinct areas.

(II) There are *no* volume functions! There is no way to way to assign to every bounded subset X of $\mathbb{R}^3$ a number $\mu(X)$ such that the rules (a)-(d) (adapted for $\mathbb{R}^3$) hold.

Actually, in 1924 the Polish mathematicians Stefan Banach and Alfred Tarski proved the following disconcerting fact, deservedly known as the Banach-Tarski paradox. It is possible to take the closed unit ball $B := \{x \in \mathbb{R}^3 : \|x\| \leq 1\}$, cut it into five pairwise disjoint subsets, move these pieces around a bit, and then reassemble them in such a way that you obtain two disjoint balls of radius 1 each. (Doesn't that knock your socks off?)

If there existed a volume function, such a feat would be impossible: The volume of the ball B would equal the sum of the volumes of the five pieces. Moving them (= applying isometries) would not change the volume; but after reassembling, the total volume would be twice the volume of B.

The Banach-Tarski paradox has been widely generalized. If physics were as easy as mathematics, you could, as it is sometimes put, cut a pool ball into finitely many pieces and use them to make a life-size (and solid) statue of Hausdorff.

Further Reading

Wagon, S., *The Banach-Tarski Paradox*, Cambridge University Press, Cambridge, 1986.

Exercises

13.A. Let (X, ω) be a topological space.
      (i) Show that the union of any two compact subsets of X is compact.

(ii) The intersection of two compact sets may fail to be compact. For an example, let X be $[0, 1]$ and let ω be the topology ω_1 described in Exercise 12.A. Show that $[0, 1)$ and $(0, 1]$ are ω_1-compact, but their intersection is not.

(iii) Show that the intersection of two compact subsets of X is compact as soon as ω is Hausdorff.

13.B. Let ω be the Euclidean topology on $\mathbb{R}$. The definition of ω can be stated as: a subset U of $\mathbb{R}$ belongs to ω if and only if for every $a \in U$, there exists an $\varepsilon > 0$ such that $(a-\varepsilon, a+\varepsilon)\backslash U$ is empty. Now, let ω' be the collection of all subsets U that have the following somewhat weaker property: For every $a \in U$, there exists an $\varepsilon > 0$ such that $(a-\varepsilon, a+\varepsilon)\backslash U$ is countable.

(i) Show that ω' is a topology, strictly stronger than ω. Show that ω' is Hausdorff but not discrete.

(ii) Show that every countable subset of $\mathbb{R}$ is ω'-closed. Deduce that all ω'-compact sets are finite.

13.C. Let X be a compact metrizable space. Use Exercise 9.E(ii) to prove that *X is homeomorphic to a closed subspace of the Hilbert Cube $[0, 1]^N$*. (See Exercise 12.C.)

(Conversely, as the Hilbert Cube itself is compact, metrizable, so are all of its closed subspaces.)

13.D. Let $(x_\tau)_{\tau \in T}$ be a net in a compact topological space X; let $a \in X$ and suppose the net does not converge to a. Show that the net has a subnet converging to a point distinct from a.

13.E. *Let ω_0 be a compact Hausdorff topology on a set X. Use Corollary 13.23 to show:*

   If ω is a topology on X that is strictly stronger than ω_0, then ω is Hausdorff but not compact. If ω is a topology on X that is strictly weaker than ω_0, then ω is compact but not Hausdorff.

13.F. Show that the restriction of a Hausdorff topology must be Hausdorff.

13.G. Let ω be as in Exercise 11.B. Show that ω is Hausdorff, not discrete, and that every ω-compact subset of $\mathbb{R}$ is finite.

14

Products and Quotients

CHAPTER

14.1

In the past, we always dealt with a set that a priori carried a topology. In this chapter, we consider two situations where a set is given a topology that is natural under the circumstances.

(I) Suppose we are given a family $(Y_i, f_i)_{i \in I}$ of pairs, each consisting of a topological space Y_i and a map $f_i : X \to Y_i$. A topology on X is considered suitable if it makes every f_i continuous. Such topologies always exist. Indeed, it is clear that the discrete topology is suitable; under it, every map of X into any topological space will be continuous. In general, if a certain topology is suitable, then so is every stronger topology. We will see (Theorem 14.4) that there exists a topology ω on X such that the suitable topologies are precisely the topologies stronger than ω.

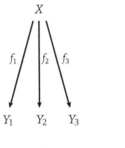

Situation (I)

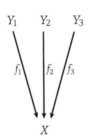

Situation (II)

As an example, let X be a subset of a topological space Y and f the natural embedding of X into Y. If our family consists of only the pair (Y, f), ω turns out to be the relative topology [Example 14.6(i)].

(II) Now suppose we have a family of pairs $(Y_i, f_i)_{i \in I}$ where each Y_i is a topological space and f_i a map from Y_i into X. Again, we may call a topology on X suitable if it makes every f_i continuous. This time the *indiscrete* topology $\{X, \varnothing\}$ certainly is suitable. Every sufficiently weak topology will be suitable. We will obtain a result similar to the above: There exists a topology ω on X such that the suitable topologies are precisely the topologies *weaker* than ω.

Product Spaces

14.2

Situation (I) is really quite simple. X is a set and I is a set; for every $i \in I$, there are given a topological space Y_i and a map $f_i : X \to Y_i$.

Let ω be any topology on X. The statement

$$\text{every } f_i \text{ is } \omega\text{-continuous} \tag{1}$$

clearly is equivalent to

for every i and every open $U \subset Y_i$,

$f_i^{-1}(U)$ belongs to ω.

Thus, if φ is the collection of sets

$$\varphi := \{f_i^{-1}(U) : i \in I; U \subset Y_i \text{ open}\},$$

then (1) is equivalent to: $\varphi \sqsubseteq \omega$.

14.3

Let X, I, Y_i, f_i, and φ be as above. The topology ω_p on X generated by φ is called the *weak topology* generated by the family $(Y_i, f_i)_{i \in I}$.

From 14.2 and 14.3 we infer:

Theorem 14.4

Let X be a set. Let I be a set; for every $i \in I$, let Y_i be a topological space and f_i a map from X into Y_i. Let ω_p be the weak topology on X generated by $(Y_i, f_i)_{i \in I}$. Then, ω_p is the weakest topology on X for which every f_i is continuous. More than that: If ω is any topology on X, then

$$\text{every } f_i \text{ is } \omega\text{-continuous} \iff \omega \text{ is stronger than } \omega_p.$$

The above description of ω_p is not particularly enlightening. However:

Theorem 14.5

Let X, I, Y_i, f_i, and ω_p be as in Theorem 14.4. If $(x_\tau)_{\tau \in T}$ is a net in X and $x \in X$, then

$$x_\tau \xrightarrow{\omega_p} x \iff \text{for every } i \in I,$$
$$f_i(x_\tau) \to f_i(x) \text{ in } Y_i.$$

Proof

Using Theorem 12.14 we see that statements (α)-(ϵ) are equivalent:

(α) $x_\tau \to x$ in the sense of ω_p.
(β) If $W \sqsubseteq \varphi$ (φ as in 14.2) and $x \in W$, then $x_\tau \in W$ for large τ.
(γ) If $i \in I$, $U \subset Y_i$ is open, and $x \in f_i^{-1}(U)$, then $x_\tau \in f_i^{-1}(U)$ for large τ.
(δ) For all i, if $U \subset Y_i$ is open and $f_i(x) \in U$, then $f_i(x_\tau) \in U$ for large τ. ∎
(ϵ) For every i, $f_i(x_\tau) \to f_i(x)$ in the topology of Y_i.

Examples 14.6

(i) Let Y be a topological space. Let $X \subset Y$ and let $f : X \to Y$ be the map $x \mapsto x$ ($x \in X$). The (trivial) family (Y, f) induces a weak toplogy ω_p on X. A subbase for ω_p is

$$\varphi := \{f^{-1}(U) : U \subset Y \text{ open}\} = \{U \cap X : U \subset Y \text{ open}\}.$$

This φ is already a topology, viz. the relative topology (12.20). Thus, ω_p is the relative topology.

(ii) Let S be a set. Let X be any set of functions $S \to \mathbb{R}$. Every element s of S determines a function $\delta_s : X \to \mathbb{R}$ by

$$\delta_s(f) := f(s) \qquad (f \in X).$$

The family $(\mathbb{R}, \delta_s)_{s \in S}$ induces a weak topology ω_p on X. If $(f_\tau)_{\tau \in T}$ is a net in X and $f \in X$, then by Theorem 14.5 we have

$$f_\tau \xrightarrow{\omega_p} f \iff f_\tau(s) \to f(s) \text{ for every } s \in S.$$

Apparently, the weak topology is the topology of pointwise convergence (Example 12.15).

14.7

Let X_1 and X_2 be topological spaces. We form their Cartesian product $X_1 \times X_2$, the set of all pairs $x = (x_1, x_2)$ with $x_1 \in X_1$ and $x_2 \in X_2$. We have coordinate maps $p_1 : X_1 \times X_2 \to X_1$ and $p_2 : X_1 \times X_2 \to X_2$, given by

$$p_1(x) := x_1, \quad p_2(x) := x_2 \qquad (x = (x_1, x_2) \in X_1 \times X_2).$$

The weak topology ω_p on $X_1 \times X_2$ determined by the family $(X_i, p_i)_{i \in \{1,2\}}$ is known as the *product topology*; it is *the weakest topology that makes both coordinate maps continuous*.

14.8

It follows from Theorem 14.5 that ω_p-convergence is "coordinatewise convergence": for a net $(x_\tau)_{\tau \in T}$ in $X_1 \times X_2$ and an element x of $X_1 \times X_2$ we have

$$x_\tau \xrightarrow{\omega_p} x \iff (x_\tau)_1 \to x_1 \text{ and } (x_\tau)_2 \to x_2.$$

Hence, if Y_1, Y_2 are topological spaces and X_i is homeomorphic to Y_i $(i = 1, 2)$, then $X_1 \times X_2$ is homeomorphic to $Y_1 \times Y_2$. More generally, a pair of continuous maps $X_1 \to Y_1$ and $X_2 \to Y_2$ determines in a natural way a continuous map $X_1 \times X_2 \to Y_1 \times Y_2$.

A subbase λ for the product topology on $X_1 \times X_2$ is formed by the sets $p_1^{-1}(U)$ ($U \subset X_1$ open) and $p_2^{-1}(V)$ ($V \subset X_2$ open), i.e., the sets $U \times X_2$ ($U \subset X_1$ open) and $X_1 \times V$ ($V \subset X_2$ open).

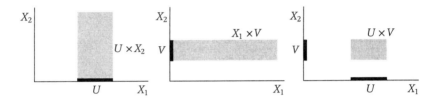

The collection $\lambda^\cap$ (terminology of Theorem 12.13) is easily seen to be

$$\{U \times V : U \subset X_1 \text{ open}, \ V \subset X_2 \text{ open}\}.$$

Thus, these "rectangles" $U \times V$ form a base for the product topology.

Examples 14.9

(i) The product topology on $\mathbb{R} \times \mathbb{R}$ is just the Euclidean topology on $\mathbb{R}^2$. (Consider coordinatewise convergence.)

(ii) Take a subset A of the open upper half-plane $\{(x, y) \in \mathbb{R}^2 : y > 0\}$. If we identify $\mathbb{R}^2$ with the plane $\{(x, y, z) \in \mathbb{R}^3 : z = 0\}$, then by rotating A around the x-axis we obtain a subset X of $\mathbb{R}^3$:

$$X := \{(x, y, z) \in \mathbb{R}^3 : (x, \sqrt{y^2 + z^2}) \in A\}.$$

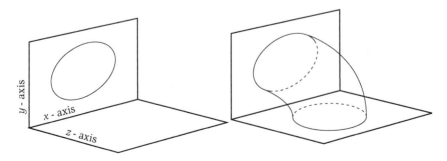

Let Γ be the unit circle

$$\Gamma := \{(u, v) \in \mathbb{R}^2 : u^2 + v^2 = 1\}.$$

The formula

$$\big((x, y), (u, v)\big) \longmapsto (x, yu, yv)$$

defines a bijection $A \times \Gamma \to X$ that is easily seen to be a homeomorphism.

(iii) For instance, let A be a circle. Then X is shaped like a tire or (the surface of) a donut. (See the picture on the previous page.) As A is homeomorphic to Γ, X is homeomorphic to $\Gamma \times \Gamma$.

A topological space homeomorphic to $\Gamma \times \Gamma$ is called a *torus*. Product topological spaces inherit many properties from their factors:

Theorem 14.10
Let X_1, X_2 be topological spaces. If X_1 and X_2 are Hausdorff, then so is $X_1 \times X_2$. If X_1 and X_2 are compact, then so is $X_1 \times X_2$.

Proof
Suppose X_1 and X_2 are Hausdorff. Take $x = (x_1, x_2)$ and $y = (y_1, y_2)$ in $X_1 \times X_2$ with $x \neq y$. Either $x_1 \neq y_1$ or $x_2 \neq y_2$; let us assume the former. There exist disjoint open subsets U and V of X_1 with $x_1 \in U$ and $y_1 \in V$. Then $U \times X_2$ and $V \times X_2$ are disjoint open subsets of $X_1 \times X_2$, and $x \in U \times X_2$ and $y \in V \times X_2$.

Now suppose X_1 and X_2 are compact. Let $(x_\tau)_{\tau \in T}$ be a net in $X_1 \times X_2$; it suffices to prove the existence of a convergent subnet. Every x_τ is a pair $\big((x_\tau)_1, (x_\tau)_2\big)$. In X_1, the net $\big((x_\tau)_1\big)_{\tau \in T}$ has a subnet $\big((x_{\alpha(\sigma)})_1\big)_{\sigma \in S}$ converging to some element a_1 of X_1. In X_2, the net $\big((x_{\alpha(\sigma)})_2\big)_{\sigma \in S}$ has a subnet $\big((x_{\alpha(\beta(\rho))})_2\big)_{\rho \in R}$ converging to some $a_2 \in X_2$. Then, $\big(x_{\alpha(\beta(\rho))}\big)_{\rho \in R}$ is a subnet of $(x_\tau)_{\tau \in T}$ and converges (coordinatewise) to (a_1, a_2). ∎

Quotient Spaces

14.11
Now let us turn to the Situation (II) of 14.1. We work with a set X, a family $(Y_i)_{i \in I}$ of topological spaces, and for each i a map $f_i : Y_i \to X$. For a topology ω on X, we see that all maps f_i are continuous if and only if

$$U \in \omega \implies \text{for every } i, f_i^{-1}(U) \text{ is open in } Y_i,$$

i.e., if and only if ω is a subcollection of

$$\omega_q := \{U : U \subset X; \text{ for every } i, f_i^{-1}(U) \text{ is open in } Y_i\}.$$

Now ω_q happens to be a topology. Indeed, it is easy to see that ω_q satisfies T1 and T3 of 11.12 or 12.1. We sketch a proof of T2. Let $\varphi \sqsubset \omega_q$; we prove $\bigcup \varphi \in \omega_q$. Take $i \in I$; we have to show that $f_i^{-1}(\bigcup \varphi)$ is open in Y_i. But $f_i^{-1}(\bigcup \varphi)$ is the union of all sets $f_i^{-1}(U)$ with $U \in \varphi$, and these sets are open in Y_i because $\varphi \sqsubset \omega_q$.

Thus, we have a topology ω_q on X, the strongest for which every f_i is continuous.

ω_q-convergence is not as easily described as ω_p-convergence. On the other hand:

Theorem 14.12
Let X, $(Y_i, f_i)_{i \in I}$ and ω_q be as above. Let g be a map of X into a topological space Z. Then, g is ω_q-continuous if and only if every $g \circ f_i$ is continuous $X_i \to Z$.

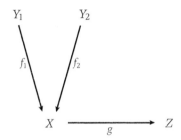

Proof
Of course, if g is continuous, then so is every $g \circ f_i$. Now assume that $g \circ f_i$ is continuous for every i. Let W be an open subset of Z and let $U := g^{-1}(W)$. For every i, $f_i^{-1}(U) = (g \circ f_i)^{-1}(W)$, so $f_i^{-1}(U)$ is open in Y_i. Then $U \in \omega_q$. Consequently, g is continuous. ∎

Example 14.13
Let X be $\mathbb{R}^2$. Every real number c determines two maps, f_c and f^c of $\mathbb{R}$ into $\mathbb{R}^2$:

$$f_c(x) := (x, c), \quad f^c(x) := (c, x) \qquad (x \in \mathbb{R}).$$

The family of all pairs $(\mathbb{R}, f_c)$ and $(\mathbb{R}, f^c)$ defines a topology ω_q on $\mathbb{R}^2$. A function $g : \mathbb{R}^2 \to \mathbb{R}$ is continuous relative to ω_q if and only if for every c, the functions

$$x \longmapsto g(x, c) \quad \text{and} \quad x \longmapsto g(c, x)$$

are continuous; i.e., if and only if g is separately continuous in the sense of Exercise 3.H and Example 10.3(ii). This brings separate continuity under the umbrella of Topology.

14.14

Mostly, in Situation (II), the family $(Y_i, f_i)_{i \in I}$ contains only one pair (Y, f) and f is surjective:

Let Y be a topological space and f a map of Y onto a set X. Then

$$\omega_q := \{U : U \subset X; f^{-1}(U) \text{ is open in } Y\}$$

is a topology on X, the so-called *quotient topology* induced by Y and f. It is the strongest topology on X that makes f continuous.

A map g of X into a topological space Z is continuous relative to ω_q if and only if the composite map $g \circ f$ of Y into Z is continuous.

14.15

For an intuitive understanding of the quotient topology, we return to some constructions we discussed in the exercises of Chapter 1. Again, take a rectangular strip of paper:

You can form a ring by gluing the ends AB and $A'B'$ together, or a "Moebius strip" if you first give the strip a twist:

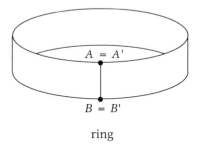

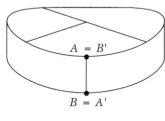

ring   Moebius strip

In a readily understood symbolism,

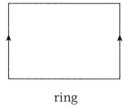

ring   Moebius strip

It is easy to give a mathematical description of a subset of $\mathbb{R}^3$ that "looks like" a ring, e.g.,

$$\{(x, y, z) \in \mathbb{R}^3 : x^2 + y^2 = 1, \ -1 \leq z \leq 1\}.$$

It is considerably harder to describe a Moebius strip in similar terms. However, the difficulties fall away if we agree that we are not interested in its geometrical but only in its topological properties. By using the quotient topology, we can construct a topological space that merits the name "Moebius strip."

Mathematically speaking, our piece of paper is a rectangle in $\mathbb{R}^2$, the unit square $S = [-1, 1] \times [-1, 1]$, say. After twisting and gluing, we end up with some subset M of $\mathbb{R}^3$. We may not know precisely what M is, but we do know that our manipulations result in a (sequentially) continuous surjection $F : S \to M$. This F is bijective except that for every $y \in [-1, 1]$, the points $(-1, y)$ and $(1, -y)$ get the same image in M. Surprisingly, that is all the information we need. It completely determines the topological structure of M and even enables us to construct a space that is homeomorphic to M.

Note that the restriction of F to the subset $(-1, 1] \times [-1, 1]$ of S is a bona fide bijection $g : (-1, 1] \times [-1, 1] \to M$. Now, $g^{-1} \circ F$ is a map $f : S \to (-1, 1] \times [-1, 1]$:

$$f(x, y) = (x, y) \quad \text{for } x \in (-1, 1], y \in [-1, 1],$$
$$f(-1, y) = (1, -y) \ \text{for } y \in [-1, 1]. \tag{$*$}$$

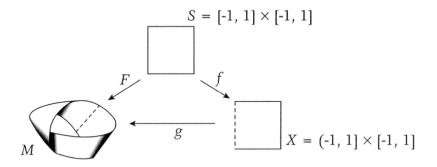

We do not know M, F, or g, but we know f, and that is enough for our purposes. Indeed, let X be the set $(-1, 1] \times [-1, 1]$ endowed with the quotient topology induced by f. As $g \circ f$ ($= F$) is continuous, so is $g : X \to M$. Furthermore, X is compact, being the image of S under a continuous map, and M is Hausdorff. Hence, g must be a homeomorphism (Corollary 13.23) and X is homeomorphic to the Moebius strip!

All this was done under the assumption that our physical manipulations with the strip of paper can be given a mathematical meaning, but our next step is obvious: We *define* a Moebius strip to be any topological space that is homeomorphic to X, i.e., to the space $(-1, 1] \times [-1, 1]$ under the quotient topology induced by the map f given by $(*)$.

This, of course, only raises the question if we can find a Moebius strip in $\mathbb{R}^3$, so that, for instance, "the" Moebius strip will be metrizable. As a matter of fact, we can do so by a procedure that is really the above argument turned upside down.

Define $F : S \to \mathbb{R}^3$ by

$$F(x, y) := \big((2+xy) \cos \pi x, (2+xy) \sin \pi x, (1-|x|)y\big).$$

Clearly, F is continuous, and one can verify that for (x, y) and (x', y') in S,

$$F(x, y) = F(x', y') \iff f(x, y) = f(x', y').$$

It follows that

$$g : f(x, y) \longmapsto F(x, y) \qquad ((x, y) \in S)$$

is a bijection of X onto $F(S)$. By Theorem 14.12, g is continuous. Then, X being compact and $F(S)$ being Hausdorff, g is a homeomorphism and $F(S)$ is a Moebius strip!

14.16

According to the mathematical parlance, the Moebius strip is obtained from the rectangle S by "identifying" the points $(-1, y)$ and $(1, -y)$ for each $y \in [-1, 1]$. This technique is used quite frequently as a formal equivalent of gluing.

Again, let us start with a rectangle. We form a ring, or, let us say, a tube, by gluing the left and right edges together. If we bend this tube and do some more gluing, we obtain a torus-like object:

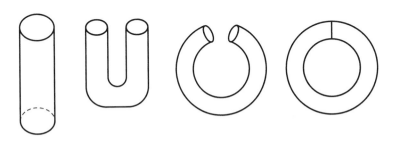

Taking for our rectangle the unit square $S = [-1, 1] \times [-1, 1]$ our manipulations amount to "identifying" $(-1, y)$ with $(1, y)$ for every $y \in [-1, 1]$ and $(x, -1)$ with $(x, 1)$ for every $x \in [-1, 1]$.

Following the ideas expounded above, we consider the set

$$S_0 := (-1, 1] \times (-1, 1];$$

we map S onto S_0 by

$$F(x, y) := (x, y) \qquad \text{if } (x, y) \in S_0,$$
$$F(-1, y) := (1, y) \qquad \text{if } -1 < y \le 1,$$
$$F(x, -1) := (x, 1) \qquad \text{if } -1 < x \le 1,$$
$$F(-1, -1) := (1, 1),$$

and we impose the quotient topology on S_0.

We expect the resulting topological space to be a torus. [See Example 14.9(iii).] Indeed, the formula

$$g : (x, y) \longmapsto \big((\cos \pi x, \sin \pi x), (\cos \pi y, \sin \pi y)\big)$$

defines a bijection $g : S_0 \to \Gamma \times \Gamma$. Following the last lines of 14.15, one easily shows g to be a homeomorphism.

Example 14.17
Things get more interesting when the quotient space is not so simple.

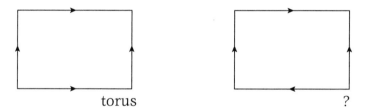

torus   ?

The diagram on the left suggests the construction of the torus out of a rectangle, as carried out in 14.16. Can we interpret the other one? Mathematically, there is no problem. In the unit square $S = [-1, 1] \times [-1, 1]$, one wishes to identify $(-1, y)$ with $(1, y)$ for $y \in [-1, 1]$ and $(x, -1)$ with $(-x, 1)$ for $x \in [-1, 1]$. This can be done as follows. Take $S_0 := (-1, 1] \times (-1, 1]$ and define a surjection $F : S \to S_0$:

$$F(x, y) := (x, y) \qquad \text{if } (x, y) \in S_0,$$
$$F(-1, y) := (1, y) \qquad \text{if } -1 < y \le 1,$$
$$F(x, -1) := (-x, 1) \qquad \text{if } -1 \le x \le 1,$$
$$F(1, -1) := (1, 1),$$

[Watch the inequality signs! We need $F(x, y) \in S_0$ for all (x, y).] Under the quotient topology induced by F, S_0 is a topological space K, the so-called *Klein bottle*.

So far, so good. But we run into trouble if we try to visualize the Klein bottle in our three-dimensional space. As we did in 14.16, we can first form the tube and then try to stick the bounding circles together:

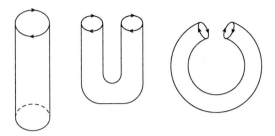

You see the problem: the directions of the circles do not match. Giving the tube a twist before bending it, the way one twists a candy wrapper,

does not help: It will not change the directions. Cheating a bit does help:

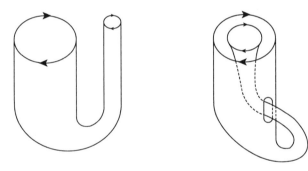

It is now easy to match the circles, and, in fact, this is the way the Klein bottle is usually pictured.

Actually, one can prove that no subset of $\mathbb{R}^3$ is homeomorphic to the Klein bottle. It can, however, be found in $\mathbb{R}^4$. Consider $f : S \to \mathbb{R}^4$ defined by

$$f(x, y) := \left((1+|x|) \cos \pi y, (1+|x|) \sin \pi y, \sin \pi x \cos \frac{\pi y}{2}, \sin \pi x \sin \frac{\pi y}{2} \right).$$

This f is continuous, and for $(x, y), (x', y') \in S$, one can verify that

$$f(x, y) = f(x', y') \iff F(x, y) = F(x', y').$$

Then $F(x, y) \longmapsto f(x, y)$ is a bijection of K onto $f(S)$, which, as before, is a homeomorphism.

We have seen that the Klein bottle can be obtained from a cylinder

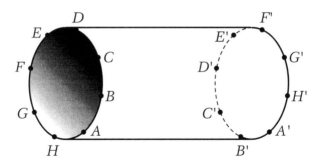

by identifying A with A', B with B', and so on. On the cylinder, draw straight lines from C to G' and from G to C'. On the Klein bottle, C is the same point as C', and G is G', so that the two lines together form a loop, a topological circle. If we cut the bottle along this loop, we get two pieces:

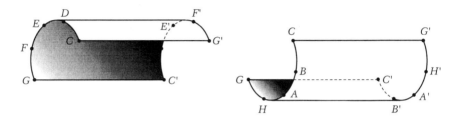

As you see, each of these pieces actually is a Moebius strip! Thus, you can get a Klein bottle by taking two Moebius strips and gluing their edges together. (Note that the edge of a Moebius strip is indeed a topological circle.)

> A mathematician named Klein
> Thought the Moebius strip was divine.
> He said: "If you glue
> The edges of two,
> You'll get a weird bottle like mine."

Extra: Surfaces

Consider the following topological spaces:

Group I
- a straight line;
- the graph of the function $x \mapsto |x|$ $(x \in \mathbb{R})$;
- an infinite helix;
- a circle;

Group II $\begin{cases} \text{- a plane;} \\ \text{- an infinite cylinder;} \\ \text{- a sphere;} \\ \text{- a torus.} \end{cases}$

It will be clear what the difference is between the spaces of Group I and those of Group II. Mathematically, the spaces of Group I are one-dimensional, the other ones two-dimensional.

If A is one of the spaces of the first group and a is any point of A, then in the neighborhood (in a nonmathematical sense) of a, the space A looks like an open interval. It may be bent or kinked, but, topologically, it is an interval. We can formally define a topological space A as one-*dimensional* if every point of A is contained in an open subset of A that is homeomorphic to an open interval (or to $\mathbb{R}$, if you prefer). A space that looks like the figure

is not one-dimensional in this sense; neither is the interval $[0, 1]$. Similarly, a space is two-*dimensional* if every point is contained in an open set that is homeomorphic to an open disk in the plane, or, equivalently, to $\mathbb{R}^2$. In the same way, one can define n-dimensionality for $n \geq 3$.

There are many variations of these definitions in the literature. Some of them allow dimensions that are not integers. According to the one due to Hausdorff, the Cantor Set (see Example 6.10) has dimension $1/3$; here, we are in the realm of "fractals," which belongs to geometry rather than topology. We are not going to worry about those but stick to the notions we have just introduced.

One-dimensional spaces are also called *curves*, two-dimensional ones *surfaces*. (As is often the case, the terminology is not unambiguously fixed. You will notice that the word "curve" here has another meaning than in the main text of the book.) The spaces of Groups I and II, above, are easily recognized as curves and surfaces. It is harder to see that the Klein bottle is a surface. The following picture, in the style of those in Example 14.17, helps:

(Actually, there is more symmetry in the Klein bottle than you may think. If a and b are points of the Klein bottle, there is a homeomorphism of the bottle onto itself that maps a onto b. Thus, no point of the bottle is in any topological sense special.)

Extensive studies have been made of curves and surfaces that are both connected and compact. Every connected compact curve turns out to be homeomorphic to a circle. There are many connected compact surfaces, though. We already know the sphere and the torus, and one easily visualizes donut-like objects with two, three, or more holes. (Compactness precludes the presence of infinitely many holes.) These are the so-called "orientable" (connected, compact) surfaces. Topologically speaking, there is precisely one for each number of holes, a sphere having zero holes. Apart from them there are certain "nonorientable" surfaces such as the Klein bottle and the projective plane (Exercise 14.G). All are homeomorphic to certain subsets of $\mathbb{R}^4$.

The n-dimensional spaces were first studied by the German mathematician Bernhard Riemann (1826-1866). Two-dimensional spaces played a role in his 1851 thesis on complex function theory: The graph of a continuous function $\mathbb{C} \to \mathbb{C}$ is a 2-dimensional subset of $\mathbb{C}^2$ ($= \mathbb{R}^4$). Three years later, he published his work on n-dimensional spaces, which rapidly became very influential in the development of mathematics. An extensive theory of such spaces was set up and turned out to yield precisely the geometrical background Einstein later needed for his general theory of relativity.

Exercises

14.A. Let (X_1, d_1) and (X_2, d_2) be metric spaces. Show that *the product topology on $X_1 \times X_2$ is metrizable and is, in fact, the d-topology where d is the sum-metric on $X_1 \times X_2$.* (See Exercise 5.H.)

14.B. A topological space (X, ω) is called *completely regular* if for every closed subset A of X and every $a \in X \backslash A$, there exists a continuous function f on

X such that

$$f(a) = 1; \quad f(x) = 0 \text{ for every } x \in A.$$

Let (X,ω) be completely regular. Let $C(X)$ be the set of all ω-continuous functions $X \to \mathbb{R}$. Show that ω is precisely the weak topology induced by the family $(\mathbb{R}, f)_{f \in C(X)}$.

14.C. The *graph* of a map $f : X \to Y$ is the set

$$G_f := \{(x, y) \in X \times Y : y = f(x)\}.$$

  (i) Let f be a map of a topological space X into a compact space Y and suppose the graph of f is closed in the product topology of $X \times Y$. Prove that f must be continuous. (With Exercise 13.D, show that f preserves convergence of nets.)
  (ii) To show the relevance of the compactness of Y, give an example of a noncontinuous function $[0, 1] \to \mathbb{R}$ whose graph is closed.
  (iii) Now, let f be a map of a topological space X into a Hausdorff space Y and suppose f is continuous. Show that it has a closed graph.

14.D. Let (X, ω) and (X', ω') be compact Hausdorff spaces. Let $f : X \to X'$ be surjective and ω-ω'-continuous. Prove that ω' is the quotient topology induced by X and f. (Can you use Exercise 13.E?)

14.E. (Regarding Example 14.13.) For $(a, b) \in \mathbb{R}^2$ and $\varepsilon > 0$, define

$$P_\varepsilon(a, b) := \{(x, b) : |x-a| < \varepsilon\} \cup \{(a, y) : |y-b| < \varepsilon\}.$$

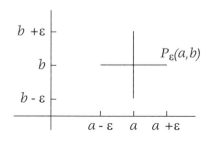

Maltese cross

We will say that a subset U of $\mathbb{R}^2$ is *P-open* if for every $(a, b) \in U$, there exists a positive ε such that $P_\varepsilon(a, b) \subset U$.
  (i) Show that the P-open sets form a topology ω_+ on $\mathbb{R}^2$.
  (ii) Show that ω_+ is precisely the topology obtained in Example 14.13.
  (iii) Show that ω_+ is stronger than the Euclidean topology and that the "Maltese cross" is P-open but not open in the Euclidean sense. The "Maltese cross" is the set of all points (x, y) for which $-1 < x < 1$, $-1 < y < 1$ and

$$|x| < \tfrac{1}{2}|y| \quad \text{or} \quad |y| < \tfrac{1}{2}|x| \quad \text{or} \quad (x, y) = (0, 0).$$

[The "Plus-plane" was discovered by the (sadly ignored) Roman mathematician Gaius Plus Minus (46B.C-A.D.19) See the footnote to page 219 of Morris Kline's *Why the Professor Can't Teach*, St. Martin's Press, New York, 1977.]

14.F. (i) If we take a piece of string and tie the ends together we get a loop.

Somewhat more scientifically: If in the unit interval $[-1, 1]$, we "identify" -1 with 1, we obtain a space that is homeomorphic to a circle.
Give an explicit example of a map F of $[-1, 1]$ onto the unit circle
$\Gamma := \{x \in \mathbb{R}^2 : \|x\| = 1\}$ that is bijective except that $F(-1) = F(1)$,
and show that the quotient topology is just the natural topology of Γ.
(ii) Show that if in the closed unit disk $\Delta := \{x \in \mathbb{R}^2 : \|x\| \leq 1\}$ we
identify all points of the boundary Γ with each other, we obtain a
space homeomorphic with the "sphere" $\{y \in \mathbb{R}^3 : \|y\| = 1\}$. [Hint:
Consider the map $F : \Delta \to \mathbb{R}^3$ defined by

$$F(x) := \left(x_1 s(\|x\|), x_2 s(\|x\|), \cos \pi \|x\| \right)$$

where s is the continuous function $\mathbb{R} \to \mathbb{R}$ with $s(t) := t^{-1} \sin \pi t$ if
$t \neq 0$, $s(0) := \pi$.]

14.G. The *projective plane* is the topological space Π obtained from the unit
square S by identifying $(-1, y)$ with $(1, -y)$ for $y \in [-1, 1]$ and $(x, -1)$
with $(-x, 1)$ for $x \in [-1, 1]$.

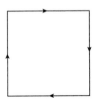

Show that there is a set in $\mathbb{R}^4$ that is homeomorphic to the projective plane.
(There is none in $\mathbb{R}^3$.) To this end, consider the map $F : S \to \mathbb{R}^4$:

$$F(x, y) := (2+xy)(\cos \pi x, \sin \pi x, \cos \pi y, \sin \pi y).$$

15

The Hahn-Tietze-Tong-Urysohn Theorems

Urysohn's Lemma

15.1

The objects of our study, topological spaces, display a great diversity. From the point of view of Analysis, one often is interested in continuous ($\mathbb{R}$-valued) functions on them. For familiar spaces such as $\mathbb{R}$ and $\mathbb{R}^2$, we are used to an abundance of continuous functions. On the other hand, there exist nontrivial topological spaces that admit no nonconstant continuous functions. For an example, see Exercise 15.H.

It is somewhat of a miracle that for a large collection of topological spaces, including the compact Hausdorff ones, there are plenty of continuous functions.

What do we mean by "plenty"? The following considerations lead to a useful interpretation of the word.

Let X be a topological space, and A and B subsets of X. We say that A and B are *separated by continuous functions* if there exists a continuous $f : X \to \mathbb{R}$ such that

$$f(x) = 1 \ \text{ for all } \ x \in A, \qquad f(x) = 0 \ \text{ for all } \ x \in B. \tag{1}$$

This can be the case only if the closures of A and B are mutually disjoint. Indeed, if (1) holds (and f is continuous), then the closed set $f^{-1}(\{1\})$ contains A, hence contains A^-, so $f = 1$ everywhere on A^-; similarly, $f = 0$ everywhere on B^-.

We say that X "admits plenty of continuous functions" if all sets that could possibly be separated by continuous functions actually are. More precisely, the condition is:

> if $A, B \subset X$ and $A^- \cap B^- = \varnothing$, then
>
> A and B are separated by continuous functions.

This condition clearly amounts to the same as

> if A, B are disjoint closed subsets of X, then
>
> A and B are separated by continuous functions.

Metrizable spaces admit plenty of continuous functions. For a proof, suppose A and B are disjoint closed sets in a metrizable space X. Without restriction, assume that neither is empty. Choose a metric for the topology of X. For every $x \in X$, we denote by $f_A(x)$ and $f_B(x)$ the distances from x to A and B, respectively. (See Exercises 5.K and 6.E.) Then,

$$x \longmapsto \frac{f_B(x)}{f_A(x) + f_B(x)}$$

is a continuous function that is 1 everywhere on A and 0 everywhere on B.

We have already hinted that compact Hausdorff spaces admit plenty of continuous functions, too. That is surprising, because, unlike metric spaces, compact Hausdorff spaces have inherently nothing to do with the real-number system.

15.2

Suppose A and B are disjoint closed subsets of a topological space X. How could one go about constructing a continuous function f with $f = 1$ on A and $f = 0$ on B? In the following, we appeal to your intuition, but at the same time, we present a precise mathematical reasoning, constituting a proof of a theorem that we will pull out of the hat afterward. (Lemma 15.6).

First, a little shift of perspective. Put $A_1 := A$ and $A_0 := X \backslash B$. Then,

$$A_1 \text{ is closed}, \quad A_0 \text{ is open}, \quad A_1 \subset A_0,$$

and we try to obtain a continuous $f : X \to \mathbb{R}$ with $f = 1$ on A_1 and $f = 0$ off A_0.

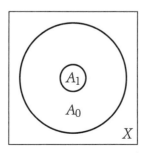

We intend to build up a function the way one makes a wedding cake, layer by layer. The top level of the cake is going to lie above the region A_1, at height 1, while outside A_0, on the rim of the plate, the cake will have height 0. Our construction takes infinitely many steps, the idea being to make an increasing sequence $f_0, f_1, f_2, \ldots$ of functions that become, as it were, more and more continuous. (We do not recommend that you actually use the complete recipe for the next wedding.)

We start out from a cake of height 1 on A_1 but 0 elsewhere:

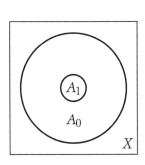
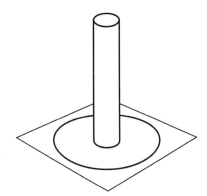

The corresponding function f_0 on X is given by

$$f_0(x) = \begin{bmatrix} 1 & \text{if} & x \in A_1, \\ 0 & \text{if} & x \notin A_1. \end{bmatrix}$$

As our first step, we choose a set $A_{1/2}$ "between" A_1 and A_0; above the new region $A_{1/2} \backslash A_1$, we give the cake height $\frac{1}{2}$:

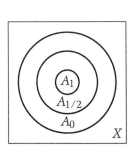
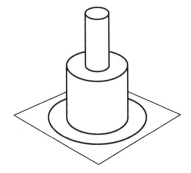

The height of the cake is now described by the function f_1:

$$f_1(x) = \begin{bmatrix} 1 & \text{if } x \in A_1, \\ \frac{1}{2} & \text{if } x \in A_{1/2} \backslash A_1, \\ 0 & \text{if } x \notin A_{1/2}. \end{bmatrix}$$

This kind of interpolation is typical for our construction. In the second stage, we choose sets $A_{3/4}$ and $A_{1/4}$ with

$$A_1 \subset A_{3/4} \subset A_{1/2} \subset A_{1/4} \subset A_0;$$

we make a four-layer cake:

   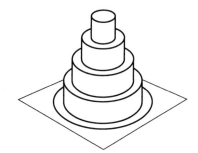

and a function f_2:

$$f_2(x) = \begin{bmatrix} 1 & \text{if } x \in A_1, \\ k/4 & \text{if } x \in A_{k/4} \backslash A_{(k+1)/4} \quad (k = 1, 2, 3), \\ 0 & \text{if } x \notin A_{1/4}. \end{bmatrix}$$

Probably, with a real wedding cake one would stop here. The mathematical construction, however, can be continued indefinitely. At stage n, we obtain $2^n + 1$ sets

$$A_1 \subset A_{(2^n - 1)2^{-n}} \subset \cdots \subset A_{2.2^{-n}} \subset A_{2^{-n}} \subset A_0 \qquad (*)$$

and a function f_n given by

$$f_n(x) = \begin{bmatrix} 1 & \text{if } x \in A_1, \\ k.2^{-n} & \text{if } x \in A_{k2^{-n}} \backslash A_{(k+1)2^{-n}} \quad (k = 1, \ldots, 2^n - 1) \\ 0 & \text{if } x \notin A_{2^{-n}}. \end{bmatrix}$$

It is elementary that $f_0 \leq f_1 \leq f_2 \leq \cdots$, that the sequence converges pointwise to some function $f : X \to [0, 1]$, and that

$$f_n \leq f \leq f_n + 2^{-n} \quad \text{for every } n.$$

(Thus, the convergence is even uniform.) Obviously, $f = 1$ on A_1 and $f = 0$ on $X \backslash A_0$.

15.3

So far, we have ignored all topological matters and, of course, there is absolutely no reason for f to be continuous. To ensure continuity during the construction, we will have to avoid the eventuality sketched next:

 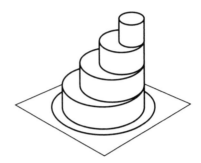

The question seems to be whether we can keep the boundaries of the sets mentioned in (∗) mutually disjoint.

For $Y, Z \subset X$, write $Y \ll Z$ in the case

$$Y \subset Z \quad \text{and} \quad \partial Y \cap \partial Z = \varnothing.$$

One easily sees that

$$Y \ll Z \quad \Longleftrightarrow \quad Y^- \subset Z^\circ.$$

The relation $\ll$ is transitive: If $Y_1 \ll Y_2$ and $Y_2 \ll Y_3$, then $Y_1 \ll Y_3$.

We want to have

$$A_1 \ll A_0, \tag{0}$$

$$A_1 \ll A_{1/2} \ll A_0, \tag{1}$$

$$A_1 \ll A_{3/4} \ll A_{1/2} \ll A_{1/4} \ll A_0, \tag{2}$$

and so on.

Formula (0) is correct, as A_1 is closed and A_0 is open. The possibility to pass from (0) to (1) and then on to (2), (3), ... is the following nontrivial property of the topological space X:

$$\text{if } Y_1, Y_2 \subset X \text{ and } Y_1 \ll Y_2, \text{ there exists}$$
$$\text{a } Z \subset X \text{ such that } Y_1 \ll Z \ll Y_2. \tag{∗}$$

This property is called normality. A more conventional (also more convenient) description follows.

15.4

A topological space X is *normal* if for any two disjoint closed subsets A and B, there exist disjoint open sets U and V such that $A \subset U$ and $B \subset V$.

This property is equivalent to (∗) of 15.3; see Exercise 15.A.

Normality follows from the existence of "plenty" of continuous functions. [Take a continuous f with $f = 1$ on A, $f = 0$ on B and let $U := \{x : f(x) > \frac{1}{2}\}$ and $V := \{x : f(x) < \frac{1}{2}\}$.] We proceed to show that the reverse is also true.

15.5

We continue our construction 15.3, assuming X to be normal, hence assuming (∗).

Starting from the valid formula (0), by normality we can choose $A_{1/2}, A_{3/4}, A_{1/4}, \ldots$ so as to ensure the truth of (1), (2), $\ldots$ At, say, stage 3 of our construction, we have

$$A_1 \ll A_{7/8} \ll A_{6/8} \ll \cdots \ll A_{2/8} \ll A_{1/8} \ll A_0$$

and the function f_3 is given by

$$f_3(x) = \begin{cases} 1 & \text{if } x \in A_1, \\ k/8 & \text{if } x \in A_{k/8} \backslash A_{(k+1)/8} \quad (k = 1, \ldots, 7), \\ 0 & \text{if } x \notin A_{1/8}. \end{cases}$$

In other words,

$$f_3 = \frac{1}{8} \sum_{k=1}^{8} \mathbf{1}_{A_{k/8}}.$$

Take $a \in X$. For at least seven values of k, we have $a \notin \partial A_{k/8}$. For such k, the indicator of $A_{k/8}$ is constant on some open neighborhood of a (viz. $A_{k/8}^{\circ}$ or $X \backslash A_{k/8}^{-}$.) It follows that there exists an open neighborhood U of a with

$$x \in U \implies |f_3(x) - f_3(a)| \le \frac{1}{8}.$$

More generally, for every $n \in \mathbb{N}$ there is an open neighborhood U_n of a such that

$$x \in U_n \implies |f_n(x) - f_n(a)| \le 2^{-n}.$$

In view of the inequality $f_n \le f \le f_n + 2^{-n}$ (last lines of 15.2), we see that

$$x \in U_n \implies |f(x) - f(a)| \le 3.2^{-n}.$$

Consequently, f is continuous.

The net result of all this is:

Theorem 15.6 (Urysohn's Lemma)

Let X be a normal topological space. Let A and B be disjoint closed subsets of X. Then, there exists a continuous function $f : X \to [0,1]$ such that $f = 1$ everywhere on A and $f = 0$ everywhere on B.

Thus, normality implies the existence of "plenty" of continuous functions and vice versa. (See the beginning of 15.5.)

Theorem 15.7
Metrizable spaces and compact Hausdorff spaces are normal.

Proof
For metrizable spaces, we have normality because disjoint closed sets are separated by continuous functions.

Suppose X is a compact Hausdorff space. Let A and B be disjoint closed subsets of X. For all $x \in A$ and $y \in B$, choose open sets U_{xy} and V_{xy} with

$$x \in U_{xy}, \quad y \in V_{xy}, \quad U_{xy} \cap V_{xy} = \varnothing.$$

Let $y \in B$. The sets U_{xy} ($x \in A$) cover A, and as a closed subset of a compact space, A is compact [Theorem 13.16(i)]. Hence, there exist $x_1, \ldots, x_N \in A$ with

$$A \subset U_y := U_{x_1 y} \cup \cdots \cup U_{x_N y}.$$

Setting

$$V_y := V_{x_1 y} \cap \cdots \cap V_{x_N y}$$

we have open sets U_y and V_y for which

$$A \subset U_y, \quad y \in V_y, \quad U_y \cap V_y = \varnothing.$$

This can be done for every $y \in B$. The sets V_y cover the compact set B, so there exist $y_1, \ldots, y_M$ such that

$$B \subset V := V_{y_1} \cup \cdots \cup V_{y_M}.$$

Put

$$U := U_{y_1} \cap \cdots \cap U_{y_M}.$$

Then, U and V are open, $A \subset U$, $B \subset V$, and $U \cap V = \varnothing$. ∎

15.8
There are many terms denoting special types of topological spaces. Some, such as "compact" and "Hausdorff," are commonplace. "Normal" is not one of those. Normally (if the word may be used here), the present authors do not give in to the temptation to introduce such relatively obscure terms. For the purposes of this book, the definition of normality is not important. The main facts are contained in Theorems 15.7 and 15.6: Metrizable spaces and compact Hausdorff spaces are normal, and normal spaces carry enough continuous functions to separate disjoint closed sets.

Interpolation and Extension

15.9
The condition (∗) of 15.3 that made the proof of Urysohn's Lemma work is of a sandwich type: Between any two sets of certain characteristics, there is another set with some special property. Many consequences of Urysohn's Lemma follow more easily from a similar theorem that is of sandwich type itself. For its formulation, we need the concept of semicontinuity:

15.10
Let X be a topological space. A function $f : X \to \mathbb{R}$ is called *lower semicontinuous* if for all $t \in \mathbb{R}$, the set

$$\{x \in X : f(x) \le t\}$$

is closed. $f : X \to \mathbb{R}$ is called *upper semicontinuous* if for all $t \in \mathbb{R}$,

$$\{x \in X : f(x) \ge t\}$$

is closed.

lower semicontinuous   upper semicontinuous

We collect a few simple observations on semicontinuity in the next theorem. The proofs are left to the reader.

Theorem 15.11
Let X be a topological space.

(i) *Let $A \subset X$. Then, $\mathbf{1}_A$ is lower semicontinuous if and only if A is open, and $\mathbf{1}_A$ is upper semicontinuous if and only if A is closed.*

(ii) *Let $f:X \to \mathbb{R}$. f is lower semicontinuous if and only if for every $t \in \mathbb{R}$, the set*

$$\{x : f(x) > t\}$$

is open. f is upper semicontinuous if and only if for every $t \in \mathbb{R}$,

$$\{x : f(x) < t\}$$

is open.

(iii) *A function $f:X \to \mathbb{R}$ is continuous if and only if it is both lower and upper semicontinuous. (Follows from (ii) and Exercise 12.H.)*

(iv) *If f and g are lower semicontinuous functions on X, then so are*

$$x \mapsto f(x) \vee g(x)$$

and

$$x \mapsto f(x) \wedge g(x).$$

(See 2.4 for $\vee$ and $\wedge$.) A similar statement holds for upper semicontinuity.

15.12

In the sketch, below, the dashed curve represents an upper semicontinuous function g, and the dotted curve, a lower semicontinuous function h. For all x, we have $g(x) \leq h(x)$.

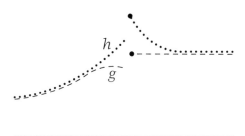

The following theorem will produce a continuous function between g and h.

Theorem 15.13 (Hahn-Tong Interpolation Theorem)
Let X be a normal topological space. Let $g:X \to \mathbb{R}$ and $h:X \to \mathbb{R}$ be upper and lower semicontinuous, respectively, and assume $g(x) \leq h(x)$ for all $x \in X$. Then, there is a continuous $f:X \to \mathbb{R}$ with $g(x) \leq f(x) \leq h(x)$ for all x.

Proof
It will be convenient to have some simple notations at our disposal. Let j and k be functions on X. We write $j \leq k$ if $j(x) \leq k(x)$ for all $x \in X$. We define functions $j \vee k$ and $j \wedge k$ by

$$j \vee k : x \longmapsto j(x) \vee k(x) \qquad (x \in X),$$

$$j \wedge k : x \longmapsto j(x) \wedge k(x) \qquad (x \in X).$$

For functions $j_0, \ldots, j_N$ on X, we let

$$\max_n j_n$$

denote the function

$$x \longmapsto j_0(x) \vee j_1(x) \vee \cdots \vee j_N(x) \qquad (x \in X).$$

Now we turn to the proof itself. Let $\mathbf{0}$ be the zero function on X.

(I) We may assume that $0 \leq g \leq h \leq 1$. Indeed, choose a strictly increasing continuous bijection $\varphi : \mathbb{R} \to (0, 1)$, e.g.,

$$\varphi(t) := \tfrac{1}{2} + \tfrac{1}{\pi} \tan^{-1}(t) \qquad (t \in \mathbb{R}).$$

Then $\varphi \circ g$ and $\varphi \circ h$ have the same semicontinuity properties as g and h and $0 \leq \varphi \circ g \leq \varphi \circ h \leq 1$. Furthermore, if f_0 is a continuous function on X and $\varphi \circ g \leq f_0 \leq \varphi \circ h$, then f_0 maps X into $(0, 1)$, so $\varphi^{-1} \circ f_0$ is a continuous function and $g \leq \varphi^{-1} \circ f_0 \leq h$.

(II) (A simple special case.) Let $A \subset X$ be closed, $U \subset X$ open, and $A \subset U$. Then, $\mathbf{1}_A$ and $\mathbf{1}_U$ are upper and lower semicontinuous, respectively, and $0 \leq \mathbf{1}_A \leq \mathbf{1}_U \leq 1$. The existence of a continuous function $f : X \to [0, 1]$ with $\mathbf{1}_A \leq f \leq \mathbf{1}_U$ now is guaranteed by Urysohn's Lemma (15.6).

(III) Returning to the general case (but with the assumption $0 \leq g \leq h \leq 1$), we first show that for every $\delta > 0$, there exists a continuous function f for which

$$g - \delta\mathbf{1} \leq f \leq h + \delta\mathbf{1}.$$

Without restriction(?), suppose $\delta = N^{-1}$ for some $N \in \mathbb{N}$.

For $t \in \mathbb{R}$, we introduce the abbreviations

$$[g \geq t] := \{x \in X : g(x) \geq t\},$$

$$[h > t] := \{x \in X : h(x) > t\}.$$

We have

$$g - \delta\mathbf{1} \leq \max_n n\delta\mathbf{1}_{[g \geq n\delta]} \leq \max_n n\delta\mathbf{1}_{[h > n\delta - \delta]} \leq h + \delta\mathbf{1}. \qquad (*)$$

The second and third of these inequalities are elementary. For the first, it is enough to observe that for every $x \in X$, there exists an n with $n\delta \leq g(x) \leq n\delta + \delta$ and that then, $g(x) - \delta \leq n\delta = n\delta\mathbf{1}_{[g \geq n\delta]}(x)$.

For each n, it follows from (II) that there exists a continuous $f_n : X \to \mathbb{R}$ such that

$$\mathbf{1}_{[g \geq n\delta]} \leq f_n \leq \mathbf{1}_{[h > n\delta - \delta]};$$

with these n, we obtain from $(*)$

$$g - \delta\mathbf{1} \leq \max_n n\delta f_n \leq h + \delta\mathbf{1}.$$

Take $f := \max_n n\delta f_n$.

(IV) Let δ and f be as in (III). Then

$$g \leq g \vee (f - \delta\mathbf{1}) \leq h \wedge (f + \delta\mathbf{1}) \leq h.$$

Observe that $g' := g \vee (f - \delta\mathbf{1})$ and $h' := h \wedge (f + \delta\mathbf{1})$ are upper and lower semicontinuous, respectively [Theorem 15.11(iv)] and that $h' - g' \leq (f + \delta\mathbf{1}) - (f - \delta\mathbf{1}) = 2\delta\mathbf{1}$. Thus, for any $\delta > 0$ we can obtain an upper semicontinuous g' and a lower semicontinuous h' such that

$$\begin{bmatrix} g \le g' \le h' \le h, \\ h'-g' \le 2\delta\mathbf{1}. \end{bmatrix}$$

(V) We are almost there. Inductively, we see that there exist sequences $g_1, g_2, \ldots$ and $h_1, h_2, \ldots$ of upper and lower semicontinuous functions, respectively, such that for all n,

$$\begin{bmatrix} g \le g_1 \le g_2 \le \cdots \le g_n \le h_n \le \cdots \le h_2 \le h_1 \le h, \\ h_n-g_n \le 2^{-n}\mathbf{1}. \end{bmatrix}$$

Apparently, both sequences converge pointwise (and even uniformly) to a function f with

$$g_n \le f \le h_n \qquad (n \in \mathbb{N}).$$

Then, $g \le f \le h$. For every $t \in \mathbb{R}$, the sets $\{x : f(x) \ge t\}$ and $\{x : f(x) \le t\}$ are closed, since

$$\{x : f(x) \ge t\} = \bigcap_n \{x : g_n(x) \ge t\},$$

$$\{x : f(x) \le t\} = \bigcap_n \{x : h_n(x) \le t\}.$$

Thus, f is both upper and lower semicontinuous, hence continuous [Theorem 15.11(iii)]. ∎

15.14
We have used Urysohn's Lemma to prove the Hahn-Tong Interpolation Theorem. On the other hand, Urysohn's Lemma is a special case, as we have seen in Part (II) of the proof, above.

Theorem 15.15 (Tietze-Urysohn Extension Theorem)
Let X be a normal topological space. Let $A \subset X$ be closed and let f_0 be a continuous function on A with values in a bounded interval $[a,b]$. Then, there exists a continuous $f:X \to [a,b]$ with $f_0(x) = f(x)$ for all $x \in A$. (See also Exercise 15.B.)

Proof
Define $g, h : X \to \mathbb{R}$ by

$$g(x) := \begin{bmatrix} f_0(x) & \text{if } x \in A, \\ a & \text{if } x \notin A, \end{bmatrix} \qquad h(x) := \begin{bmatrix} f_0(x) & \text{if } x \in A, \\ b & \text{if } x \notin A. \end{bmatrix}$$

Then, $g(x) \le h(x)$ for all $x \in A$. It is not hard to verify that g is upper semicontinuous and h is lower semicontinuous. By the Hahn-Tong Theorem, there is a continuous $f : X \to \mathbb{R}$ with $g(x) \le f(x) \le h(x)$ for all x. Then,

$a \leq f(x) \leq b$ for all x, whereas for $x \in A$, we obtain $f_0(x) \leq f(x) \leq f_0(x)$, i.e., $f_0(x) = f(x)$. ∎

One important application of the Tietze-Urysohn Theorem is the proof of the Jordan Closed Curve Theorem in the next chapter. Another one is:

Theorem 15.16 (Urysohn's Metrization Theorem)
Let X be a compact Hausdorff space whose topology has a base consisting of only countably many sets. Then, X is metrizable.

[Conversely, in a compact metrizable space, the topology has a countable base; see Exercise 9.E(i).]

Proof
Let $U_1, U_2, \ldots$ be the sets belonging to a countable base for the topology of X. For certain combinations of i and j, it will happen that

$$U_i^- \cap U_j^- = \varnothing.$$

For each such pair (i, j), applying Urysohn's Lemma we choose a continuous function $f_{ij} : X \to [0, 1]$ with $f_{ij} = 1$ everywhere on U_i^- and $f_{ij} = 0$ everywhere on U_j^-. There are only countably many pairs (i, j), so we have only countably many functions f_{ij}. We arrange them into an infinite sequence

$$g_1, g_2, \ldots.$$

(Repetitions are permitted.)

If $a, b \in X$ and $a \neq b$, there is an n with $g_n(a) \neq g_n(b)$. Indeed, by Urysohn's Lemma, there exists a continuous $h : X \to [0, 1]$ with $h(a) = 1$, $h(b) = 0$. Since the sets $U_1, U_2, \ldots$ form a base for the topology, there exist i and j for which

$$a \in U_i \subset \{x : h(x) > \tfrac{2}{3}\}, \quad b \in U_j \subset \{x : h(x) < \tfrac{1}{3}\}.$$

Then, $U_i^- \cap U_j^- = \varnothing$, so in the sequence $g_1, g_2, \ldots$, there is a g_n with $g_n = 1$ on U_i^- and $g_n = 0$ on U_j^-. In particular, $g_n(a) = 1$ and $g_n(b) = 0$.

It follows that the formula

$$\delta(x, y) := \sum_{n=1}^{\infty} 2^{-n} |g_n(x) - g_n(y)| \qquad (x, y \in X)$$

defines a metric δ on X.

Now let ω be the given topology on X. For every $b \in X$, the function $x \mapsto \delta(x, b)$ is a uniform limit of ω-continuous functions, hence is itself ω-continuous. Therefore,

$$\text{if } \quad x_\tau \xrightarrow{\omega} b, \quad \text{then} \quad \delta(x_\tau, b) \to 0.$$

This means that the identity map of X is ω-δ-continuous (if the mixed notation can be excused.) Now under ω, X is compact; under the δ-topology, X is Hausdorff. Hence, by Corollary 13.23 the identity map of X is a homeomorphism relative to ω and the δ-topology, i.e., ω is the δ-topology. ∎

Another consequence of the normality of compact Hausdorff spaces is the following:

Theorem 15.17 (Baire Category Theorem; second version)
Let X be a compact Hausdorff space. If $U_1, U_2, \ldots$ are open and dense subsets of X, then their intersection is dense. (Compare Theorem 7.9.)

Proof
Let W_0 be a nonempty open set; we prove $W_0 \cap U_1 \cap U_2 \cap \cdots \neq \emptyset$.

As U_1 is dense, $W_0 \cap U_1$ contains an element x. Now, $\{x\} \ll W_0 \cap U_1$, so (by normality) there is an open set W_1 with $x \in W_1 \subset W_1^- \subset W_0 \cap U_1$. In particular, W_1 is nonempty.

Similarly (replace W_0 by W_1 and U_1 by U_2), there exists a nonempty open set W_2 for which $W_2^- \subset W_1 \cap U_2$. Inductively, one obtains a sequence $W_0, W_1, W_2, \ldots$ of nonempty open sets such that $W_n^- \subset W_{n-1} \cap U_n$ for all $N \in \mathbb{N}$. As X is compact, there is a point c of X that lies in every W_n^-. Then, $a \in W_0$ and $a \in U_n$ for all n. ∎

15.18
Things may go wrong without the Hausdorff property [but see Exercise 12.I(iv)]. For $n \in \mathbb{N}$, let $U_n := \{n, n+1, n+2, \ldots\}$. Together with the empty set, the sets $U_1, U_2, \ldots$ form a compact topology on $\mathbb{N}$; each U_n is open and dense, but their intersection is empty.

Extra: Nonstandard Mathematics

The real numbers are the primary source of inspiration for Topology and Analysis. The most familiar geometric representation of $\mathbb{R}$ is a straight line. That picture of the real numbers is helpful but not without problems. In particular, it is impossible to see what happens "far away" and the thought process of extending the segment indefinitely raises questions. For instance, is it possible that, far away on this extended line, real numbers can be found that are larger than every natural number?

The answer is "no" if we base our ideas about the real numbers on the axioms of Chapter 1. Similarly, no real positive nonzero number is closer to zero than the reciprocals of all natural numbers. This has not prevented mathematicians from thinking about infinitely large and infinitely small numbers. In fact, Leibniz phrased his inventions in Calculus in terms of such infinitesimals.

Although physicists never really discarded the intuition of the infinitely small, the foundations of Analysis, completed at the end of the nineteenth century, seemed to put their mathematical use to the rest. However, in 1962 Abraham Robinson discovered a way to revive infinitely large and infinitely small elements. He gave the system of real numbers in which such numbers figure the name

nonstandard real numbers;

in notation: $^*\mathbb{R}$, a system containing not only all real numbers but also, for every x in $\mathbb{R}$, infinitely many elements between x and (x, ∞) and between $(-\infty, x)$ and x; in addition, there are infinitely many elements to the right and to the left of $\mathbb{R}$. As one avails oneself of the complex numbers to solve, say, $x^2 = -1$, $^*\mathbb{R}$ can be used in Calculus or wherever infinitely small or infinitely large numbers appeal to the intuition. Thus, $^*\mathbb{R}$ does not simply and radically do away with $\mathbb{R}$. Rather, $^*\mathbb{R}$ is a more or less natural extension of $\mathbb{R}$.

All this would be useless if the nonstandard real numbers $^*\mathbb{R}$ would behave differently than what you are used to. They don't: $^*\mathbb{R}$ satisfies Axioms I and II of Chapter 2. It contains $\mathbb{R}$ and therefore $\mathbb{N}$. The Archimedes-Eudoxus is not just thrown overboard. What happens is roughly the following. Ordinarily, when we do Analysis, from the real number system we construct a variety of objects such as the order relation $\leq$, the sine function, or the Euclidean metric. All these objects have their counterparts in Nonstandard Analysis, the study of the nonstandard real numbers. Thus, in $^*\mathbb{R}$, there is an ordering $^*\!\leq$, extending $\leq$ in the sense that,

$$x \leq y \quad \Longrightarrow \quad x \;^*\!\leq y \quad (x, y \in \mathbb{R});$$

there is a function $^*\sin : {}^*\mathbb{R} \to {}^*\mathbb{R}$, extending the sine:

$$(^*\sin)(x) = \sin x \quad (x \in \mathbb{R}),$$

and so forth. Every subset X of $\mathbb{R}$ determines a subset $^*X$ of $^*\mathbb{R}$ with $X \subset {}^*X$ (and even $X = {}^*X \cap \mathbb{R}$).

By the so-called Transfer Principle, all valid formulas about real numbers have their star versions, valid for nonstandard numbers. For instance, for all x in $\mathbb{R}$,

$$\sin 2x = 2 \sin x \cos x;$$

the Transfer Principle guarantees that for all x in $^*\mathbb{R}$:

$$^*\sin 2x = 2^*\sin x \,{}^*\cos x.$$

(Here, we sweep a major difficulty under the carpet: What is a "formula"? A sensible answer can be given, but we are not going to do so.)

Back to our axioms. $^*\mathbb{R}$ does not precisely obey Axiom N but it obeys its star version:

$$\text{for every } x \in {}^*\mathbb{R} \text{ there is an } n \in {}^*\mathbb{N} \text{ with } x < n.$$

(We really ought to write " $^*\leq$ " instead of " $\leq$," but let's not bother.)

An element x of $^*\mathbb{R}$ is said to be infinitely large if $n \leq |x|$ for all n in $\mathbb{N}$ (not $^*\mathbb{N}$). x is called infinitely small if $|x| \leq n^{-1}$ for all n in $\mathbb{N}$.

Some simple facts:

• $^*\mathbb{N}$ contains infinitely large elements.

• If $x \in \mathbb{R}$ and $n \in {}^*\mathbb{N}$ is infinitely large, then $\frac{x}{n}$ is infinitely small.

• For every x in $^*\mathbb{R}$ that is not infinitely large, there exists a unique element y of $\mathbb{R}$ (called the standard part of x) such that $x - y$ is infinitely small; we then write $x \approx y$.

• If f is a function on $[0, 1]$, it has a nonstandard counterpart $^*f$ that is a $^*\mathbb{R}$-valued function on $^*[0, 1]$ (which is $\{x \in {}^*\mathbb{R} : 0 \leq x \leq 1\}$). This $^*f$ is an extension of f.

• f is continuous if and only if

$$x \approx y \quad \Longrightarrow \quad (^*f)(x) \approx (^*f)(y) \qquad (x, y \in {}^*[0, 1]).$$

It is the latter feature that makes nonstandard Analysis interesting for Topology. It replaces a complicated concept of continuity by one that is much simpler and appeals to the intuition.

We should mention here that nonstandard methods are not confined to the real-number system. The whole idea carries over to metric spaces and, to a lesser extent, to topological spaces.

Once the machinery has been set up, nonstandard methods sometimes give a better insight and can lead to suprisingly easy proofs for theorems about "standard" situations. As an example, we give a proof of the Intermediate Value Theorem (2-18):

Let $f : [0, 1] \to \mathbb{R}$ be continuous, and $f(0) < 0 < f(1)$. We wish to establish the existence of an $x \in [0, 1]$ with $f(x) = 0$. Clearly,

$$\text{for every } n \in \mathbb{N}, \text{ there is a } k \in \mathbb{N}$$

$$\text{with } k \leq n \text{ and } f\left(\tfrac{k-1}{n}\right) < 0 \leq f\left(\tfrac{k}{n}\right).$$

The Transfer Principle guarantees that

$$\text{for every } n \in {}^*\mathbb{N}, \text{ there is a } k \in {}^*\mathbb{N}$$

$$\text{with } k \leq n \text{ and } (^*f)\left(\tfrac{k-1}{n}\right) < 0 \leq (^*f)\left(\tfrac{k}{n}\right).$$

Take any infinitely large n in $^*\mathbb{N}$ and choose a corresponding k. As $k \leq n$, $\frac{k}{n}$ is not infinitely large and has a standard part: x.

$$f(x) = (^*f)(x) \approx (^*f)\left(\tfrac{k}{n}\right) \geq 0. \qquad (1)$$

n is infinitely large, so $\frac{k-1}{n} \approx \frac{k}{n} \approx x$ and

$$f(x) = (^*f)(x) \approx (^*f)\left(\frac{k-1}{n}\right) < 0. \tag{2}$$

From (1) and (2), it follows that $f(x) = 0$, since $f(x) \in \mathbb{R}$.

Further Reading

Hurd, A.E. and P.A. Loeb, *An Introduction to Nonstandard Real Analysis,* Academic Press, London, 1985.

Exercises

15.A. (See 15.4.) Show that a topological space X is normal if and only if it satisfies (∗) of 15.3.

15.B. Prove the following alternative version of the Tietze-Urysohn Theorem: If A is a closed subset of a normal space X and if f_0 is a continuous function on A (not necessarily bounded), then f_0 extends to a continuous function on X.

Hint: The obvious start is to apply Theorem 15.15 to the function $\tan^{-1} \circ f_0$. This will yield a continuous $g : X \to [-\pi/2, \pi/2]$, but the trouble is that g might take the values $\pi/2$ or $-\pi/2$. However, the set $B := \{x : |g(x)| = \pi/2\}$ is closed and does not meet A, furnishing another occasion to use the normality of X.

15.C. (A converse to Theorem 15.15.) Let X be a topological space. Suppose that for every closed set $A \subset X$, every bounded continuous function $A \to \mathbb{R}$ can be extended to a continuous function $X \to \mathbb{R}$. Show that X must be normal.

15.D. We have obtained the Tietze-Urysohn Extension Theorem (15.15) as an application of the Hahn-Tong Interpolation Theorem, which is a consequence of Urysohn's Lemma. The following is a more direct proof of the Tietze-Urysohn Theorem from Urysohn's Lemma:

Let A be a closed subset of a normal space X. Let f be a continuous function $A \to [0, 1]$. By Urysohn's Lemma, there is a continuous $g : X \to [0, 1]$ with

$$g = 0 \quad \text{on} \quad \{x \in A : f(x) \le \tfrac{1}{3}\},$$

$$g = 1 \quad \text{on} \quad \{x \in A : f(x) \ge \tfrac{2}{3}\}.$$

Show that $0 \le f(x) - \frac{1}{3}g(x) \le \frac{2}{3}$ for all $x \in A$. Now construct a sequence $g_1, g_2, \ldots$ of continuous functions on X such that $\sum g_n$ converges uniformly and $f(x) = \sum_{n=1}^{\infty} g_n(x)$ for $x \in A$. Thus, f has a continuous extension $X \to [0, 1]$.

15.E. (On semicontinuity.)
  (i) Let f be a lower semicontinuous function on a compact space X. Prove that f attains a smallest value.
  (ii) Let X be a topological space, F a set of lower semicontinuous function on X, and g a function on X such that

$$g(x) = \sup\{f(x) : f \in F\} \qquad (x \in X).$$

  Show that g is lower semicontinuous.
  (iii) Let X be a normal Hausdorff space (or let X be a completely regular space; see Exercise 14.B). Show that every bounded lower semicontinuous function on X is the pointwise supremum of a set of continuous functions.

15.F. *Let X be a metrizable compact space and Y a Hausdorff space such that there exists a continuous surjection $f : X \to Y$. Then Y is metrizable.* Prove that.
  Hint: By Exercise 9.E(i), the topology of X has a base φ consisting of only countably many sets. Let $\bar{\varphi}$ be the collection of all sets $U_1 \cup \cdots \cup U_N$ with $N \in \mathbb{N}$ and $U_1, \ldots, U_N \in \varphi$. This $\bar{\varphi}$ is countable and a base for the topology of X. For $U \in \bar{\varphi}$, put

$$U' := Y \backslash f(X \backslash U).$$

  Show that each U' is open in Y and that $\{U' : U \in \bar{\varphi}\}$ is a base for the topology of Y.

15.G. Prove the following theorem. *If f is a map of a compact Hausdorff space X onto a set Y such that the set*

$$\{(x, x') \in X \times X : f(x) = f(x')\}$$

is closed in the product topology of $X \times X$, then the quotient topology of Y is (compact and) Hausdorff.
  A proof can be given along these lines:
  (i) If $A \subset X$ is closed, then so is $f^{-1}(f(A))$.
  (ii) If $A \subset X$ is closed, then $f(A)$ is closed in the quotient topology.
  (iii) For every $a \in Y$, the set $\{a\}$ is closed.
  (iv) If $U \subset X$ is open, then $U' := Y \backslash f(X \backslash U)$ is open in the quotient topology.
  (v) The quotient topology is Hausdorff. (Use (iii) and the normality of X.)

15.H. If X is a topological space, we say that the continuous functions *separate the points of X* if for all $a, b \in X$ with $a \neq b$, there exists a continuous function f on X for which $f(a) \neq f(b)$. It is easily seen that such a space must be Hausdorff. The converse is false, however. There exist Hausdorff spaces whose points are not separated by continuous functions. The following is a particularly unpleasant example, one of the Frankenstein monsters of Topology.
  Let $X := \{(x, y) \in \mathbb{Q} \times \mathbb{Q} : y \geq 0\}$. For $u \in \mathbb{R}$ and $r > 0$, put

$$I_r(u) := \{(x, 0) \in X : |x - u| < r\}.$$

Let ω be the collection of all subsets U of X with the property:

For every $(a, b) \in U$, there is an $\varepsilon > 0$

such that $I_\varepsilon(a+b\sqrt{3}) \cup I_\varepsilon(a-b\sqrt{3}) \subset U$.

(i) Show that for all $u \in \mathbb{R}$ and $r > 0$, $I_r(u)$ belongs to ω, and that for all $(a, b) \in X$ and $\varepsilon > 0$, the set

$$\{(a, b)\} \cup I_\varepsilon(a+b\sqrt{3}) \cup I_\varepsilon(a-b\sqrt{3})$$

belongs to ω.

(ii) Show that ω is a topology. This is the topology we now put on X.

(iii) Show that X is a Hausdorff space.

(iv) Let $u \in \mathbb{R}$ and $r > 0$. Prove that the closure of $I_r(u)$ is

$$\{(a, b) \in X : a+b\sqrt{3} \in [u-r, u+r] \text{ or } a-b\sqrt{3} \in [u-r, u+r]\}.$$

(Make a sketch.) Deduce that for all $u, v \in \mathbb{R}$ and $r, s > 0$, the closures of the sets $I_r(u)$ and $I_s(v)$ have nonempty intersection.

(v) Use (iv) to prove:

If U, V are nonempty open sets in X, then $U^- \cap V^- \neq \varnothing$.

(vi) Now show that all continuous functions on X are constant.

16

CHAPTER

Connectedness

Basically, the notion of connectedness is very easy to grasp. In $\mathbb{R}^2$, the set

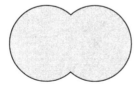

is connected; the set

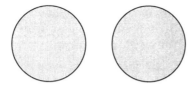

is not. There are various ways to translate this idea into mathematical terms. In this chapter, we discuss two of them. The first is the one to which mathematical usage has reserved the term "connectedness"; the second is the "path connectedness" we have already mentioned in Chapter 4. Connectedness (in the strict sense) is the more basic of the two; its definition goes back directly to the notion of a topological space. Path connectedness is intuitively more accessible but less elementary; it relies on properties of the real-number system.

Both notions and the knowledge we build up about them in the first part of this chapter will come together in our proof of the Jordan Closed Curve Theorem (16.28).

Connected Spaces

16.1

We define a topological space (X, ω) to be *connected* if $\varnothing$ and X are the only subsets of X that are both open and closed. For a subset A of X, we say that A is *connected* if (A, ω_A) is connected.

The following lemma provides a characterization of connectedness that is usually easier to apply than the above definition. We will often employ it. It provides a characterization of connected spaces in terms of continuous functions.

Lemma 16.2

A topological space X is connected if and only if every continuous function $X \to \{0,1\}$ is constant.

Proof

By Exercise 16.A, the continuous functions $X \to \{0, 1\}$ are the indicators of the subsets of X that are both closed and open. The lemma follows.   ∎

Examples 16.3

(i) Let X be $(0, 1) \cup (2, 3)$ with the usual topology. $(0, 1)$ and $(2, 3)$ are open in X, so $(0, 1)$ is open but also closed in X. Thus, X is not connected.

(ii) Let A be a subset of $\mathbb{R}$, neither $\varnothing$ nor $\mathbb{R}$. By the Connectedness Theorem 2.17, A and $X \backslash A$ have a common adherent point. It follows that A and $X \backslash A$ cannot both be closed, so that A cannot be closed and open. Therefore, $\mathbb{R}$ is connected. More generally, we have:

Theorem 16.4

$\mathbb{R}$ *and all of its intervals* $((a,b),(a,b],[a,b],(a,\infty), \dots)$ *are connected.*

Proof

By the Intermediate Value Theorem (2.18), every continuous function on an interval that takes the values 0 and 1 must also take the value $\frac{1}{2}$.   ∎

16.5

For the definition of path connectedness, we recall a bit of terminology from Chapter 4, at the same time putting it into a more general context.

A *curve* in a topological space X is a continuous map of a closed bounded interval into X. If $\varphi : [a, b] \to X$ is such a curve, then φ is said to be a curve *from $\varphi(a)$ to $\varphi(b)$* or a curve *connecting $\varphi(a)$ and $\varphi(b)$*; $\varphi(a)$ and $\varphi(b)$ are the *end points* of φ; and

$$\varphi^* := \{\varphi(t) : t \in [a, b]\}.$$

X is called *path connected* if for any two points $u, v \in X$, there exists a curve in X with end points u and v. (Then there is a curve $\varphi : [0, 1] \to X$ with $\varphi(0) = u$, $\varphi(1) = v$.)

A subset U of X is *path connected* if under the relative topology U is a path connected topological space.

$\mathbb{R}$ and all intervals are easily seen to be path connected. So are $\mathbb{R}^N$ ($N \in \mathbb{N}$) and all balls in $\mathbb{R}^N$.

Theorem 16.6
Every path connected topological space is connected.

Proof
Let X be a path connected topological space and let $f : X \to \{0, 1\}$ be continuous. Let $x, y \in X$. There exists a curve $\varphi : [0, 1] \to X$ with $\varphi(0) = x$ and $\varphi(1) = y$. Then, $f \circ \varphi$ is a continuous map $[0, 1] \to \{0, 1\}$. But by the Intermediate Value Theorem (2.18), such a map is constant and in particular

$$(f \circ \varphi)(0) = (f \circ \varphi)(1).$$

In other words, $f(x) = f(y)$. But x and y were arbitrary, so f must be constant. ∎

Thus, $\mathbb{R}^2$ and, indeed, every $\mathbb{R}^N$ is connected.

Example 16.7
Not every connected space is path connected, though. An example is furnished by the so-called "topologist's comb without zero":

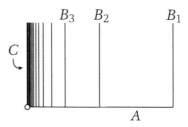

Let

$$A := \{(x, 0) : 0 < x \le 1\},$$

$$B_n := \{(\tfrac{1}{n}, y) : 0 \le y \le 1\} \quad (n \in \mathbb{N}),$$

$$C := \{(0, y) : 0 < y \le 1\},$$

$$X := A \cup \bigcup_{n=1}^{\infty} B_n \cup C.$$

[Note that $(0, 0) \notin X$.] The topology on X is Euclidean.

We first show that X is connected. Assume $f : X \to \{0, 1\}$ is continuous and $f(1, 0) = 0$. Then, $f = 0$ on the (connected!) horizontal interval A. By the same token, f is constant on each B_n, hence $f = 0$ on each B_n. Finally, for $0 < y \le 1$, we have $f(0, y) = \lim_{n \to \infty} f(\tfrac{1}{n}, y) = 0$. Thus, $f = 0$ on all of X.

This establishes the connectedness of X. We claim that X is *not* path connected. In fact, we will show that there is no curve in X with end points $(1, 0)$ and $(0, 1)$. *Suppose $\varphi : [0, 1] \to X$ is a curve, $\varphi(0) = (1, 0)$, and $\varphi(1) = (0, 1)$.* The function

$$t \mapsto \text{ first coordinate of } \varphi(t)$$

is continuous $[0, 1] \to [0, 1]$ and takes the values 0 and 1. By the Intermediate Value Theorem, for any $n \in \mathbb{N}$ the number $s_n := \tfrac{1}{2}(\tfrac{1}{n} + \tfrac{1}{n+1})$ is a value of this function. But $(s_n, 0)$ is the only point of X whose first coordinate is s_n. It follows that $(s_n, 0) \in \varphi^*$ for every n. Now, φ^* is closed in $\mathbb{R}^2$ [Exercise 9.F or 13.16(ii)] and $(s_n, 0) \to (0, 0)$, so $(0, 0) \in \varphi^*$. However, $(0, 0) \notin X$. *Contradiction.*

A continuous walk from a point inside a set A to a point outside A must hit the boundary A:

Lemma 16.8
Let A be a subset of a topological space X and $p \in A$. Let φ be a curve in X connecting p with a point outside A. Then, $\varphi^* \cap \partial A \neq \emptyset$. Moreover, there is a curve in $\varphi^* \cap A^-$ that connects p with a point of ∂A.

Proof

Without restriction, we assume that the domain of φ is $[0, 1]$ and $\varphi(0) = p$. Let

$$r := \inf\{t \in [0, 1] : \varphi(t) \notin A\}$$

and $q := \varphi(r)$. Then, $q \in \partial A$ and $t \mapsto \varphi(tr)$ $(0 \leq t \leq 1)$ is a curve in $A \cup \{q\}$, hence in $\varphi^* \cap A^-$, connecting p and q.   ∎

The following observations will be useful.

Theorem 16.9

Let X and Y be topological spaces and let $f:X \to Y$ be continuous. If $A \subset X$ is connected, then so is $f(A)$.

Proof

Let $g : f(A) \to \{0, 1\}$ be continuous. Then $g \circ f$ maps A continuously into $\{0, 1\}$, so $g \circ f$ is constant on A and g is constant on $f(A)$.   ∎

Theorem 16.10

Let X be a topological space. Then every connected subset of X has a connected closure.

Proof

Let $A \subset X$ be connected. Take a continuous function $f : A^- \to \{0, 1\}$. Then, f is constant on A, hence on A^-.   ∎

Theorem 16.11

Let α be a collection of connected subsets of a topological space and suppose there is an A_0 in α such that

$$A \cap A_0 \neq \emptyset \quad \text{for all } A \in \alpha.$$

Then $\bigcup \alpha$ is connected.

Proof

Let $f : \bigcup \alpha \to \{0, 1\}$ be continuous. For every $A \in \alpha$, f is constant on A. The given condition implies that for each A, the value of f on A equals the value of f on A_0. Hence, f is constant on $\bigcup \alpha$. ∎

16.12

We apply Theorem 16.11 in the following way.

Let X be a topological space. For each $x \in X$, let C_x be the union of all connected subsets of X that contain x. By Theorem 16.11, C_x is itself connected. Thus, C_x is the largest connected subset of X containing x.

If $x, y \in X$ and $C_x \cap C_y \neq \varnothing$, then, again by Theorem 16.11, $C_x \cup C_y$ is connected, so $C_x \cup C_y = C_x$ and $C_x = C_y$. Thus, the sets C_x form a "partition" of X:

Their union is X;

they are pairwise disjoint.

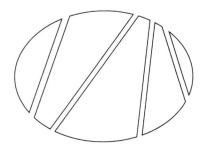

For that reason, the sets C_x are called the *components* of X. As a consequence of Theorem 16.10 we have:

Theorem 16.13

Every component of a topological space is closed.

16.14

In order to obtain similar results for path connectedness, we need the following. Let X be a topological space.

If $\varphi_1, \varphi_2 : [0, 1] \to X$ are curves and $\varphi_1(1) = \varphi_2(0)$,

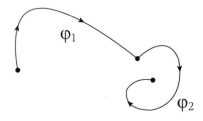

we define a curve $\varphi_1 \cdot \varphi_2 : [0, 1] \to X$ by

$$(\varphi_1 \cdot \varphi_2)(t) := \begin{cases} \varphi_1(2t) & \text{if } 0 \le t \le \frac{1}{2} \\ \varphi_2(2t-1) & \text{if } \frac{1}{2} \le t \le 1. \end{cases}$$

Then,

$$(\varphi_1 \cdot \varphi_2)^* = \varphi_1^* \cup \varphi_2^*.$$

With the aid of this observation, one easily proves the following analog of Theorem 16.8:

Theorem 16.15
Let α be a collection of path connected subsets of a topological space and suppose there is an A_0 in α such that

$$A \cap A_0 \ne \varnothing \quad \text{for all } A \in \alpha.$$

Then, $\bigcup \alpha$ is path connected.

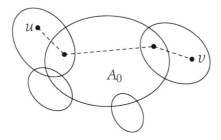

16.16
As in 16.12, for each $x \in X$ we have a largest path connected subset P_x of X containing x. These sets P_x form a partition of X and are called the *path components* of X. It follows from Theorem 16.6 that P_x is connected, so that $P_x \subset C_x$. Thus, every component of X is a union of path components of X.

Example 16.17
Path components can fail to be closed. Consider the topologist's comb without zero, described in Example 16.7. It is connected but not path connected. Hence, it has only one component but at least two path components. (How many does it have?)

From Theorem 16.15, one sees that $A \cup \bigcup_{n=1}^{\infty} B_n$ is path connected. Its closure (in the topological space X) is X itself. Thus, the analog of Theorem 16.10 falls through: The closure of a path connected set may not be path connected.

Although not every connected set is path connected, for open sets in $\mathbb{R}^n$ the two concepts are equivalent:

Theorem 16.18

Let $N \in \mathbb{N}$ and let $U \subset \mathbb{R}^N$ be open. If C is a component of U, then C is an open subset of $\mathbb{R}^N$, $\partial C \subset \partial U$, and C is path connected. In particular, if U is connected, then U is path connected.

Proof

For $a \in U$, let P_a be the path component of U containing a.

If $a \in U$, there is an $\varepsilon > 0$ such that the ball $B_\varepsilon(a) := \{x \in \mathbb{R}^N : \|x - a\| < \varepsilon\}$ is contained in U. As $B_\varepsilon(a)$ is path connected, it follows that $B_\varepsilon(a) \subset P_a$. Thus, every path component of U is open in $\mathbb{R}^N$.

C is a union of path components of U, hence is open in $\mathbb{R}^N$.

C is closed in U (see 16.12), so $C = C^- \cap U$, the bar denoting closure in the topological space $\mathbb{R}^N$. Hence, $\partial C = C^- \backslash C^\circ = C^- \backslash C = C^- \backslash U \subset U^- \backslash U = \partial U$.

To prove that C is path connected, take $a \in C$. Then, P_a is open in $\mathbb{R}^N$, hence open in C. But $C \backslash P_a$ is a union of path components of U, and therefore also open in $\mathbb{R}^N$ and in C. Thus, P_a is open and closed in C. As C is connected and $P_a \neq \varnothing$, we have $C = P_a$, i.e., C is path connected. ∎

The Jordan Theorem

We now return to a promise we made at the end of Chapter 4, to prove the Jordan Closed Curve Theorem. We claimed there that for any closed Jordan curve ω in $\mathbb{R}^2$ the set $\mathbb{R}^2 \backslash \omega^*$ is the union of two disjoint nonempty path connected sets. The following turns out to be no harder to prove.

Theorem 16.19 (Jordan Closed Curve Theorem)

Let ω be a closed Jordan curve in $\mathbb{R}^2$. Then, the set $\mathbb{R}^2 \backslash \omega^*$ has two components, one bounded and one unbounded, each of which has ω^* as its boundary.

(This version of the Jordan Theorem implies our earlier one, Theorem 4.29. Indeed, ω^* is a closed set [see Exercise 9.F and Theorem 13.16(ii)] and by Theorem 16.18, the components of $\mathbb{R}^2 \backslash \omega^*$ are path connected.)

A proof of the theorem requires considerable preparation; we conclude it in 16.28. In the meantime, we will obtain some related results of independent interest. (Theorem 16.21 and Corollary 16.22.) For the rest of this chapter, the scene is set in $\mathbb{R}^2$. Thus, U^-, U°, and ∂U will be the closure, the interior, and the boundary of U, respectively, relative to the topological space $\mathbb{R}^2$.

16.20

Before turning to the proof proper, we prepare the ground by making a few general remarks about (images of) curves and their complements.

Suppose φ is a curve in $\mathbb{R}^2$, not necessarily Jordan or closed Jordan.

(i) φ^* is a closed bounded subset of $\mathbb{R}^2$. [See Exercise 9.F and Theorem 13.16(ii).]

(ii) Hence, the components of $\mathbb{R}^2\backslash\varphi^*$ are path connected open subsets of $\mathbb{R}^2$. (Theorem 16.18).

(iii) If U is a component of $\mathbb{R}^2\backslash\varphi^*$, then $\partial U \subset \varphi^*$. Indeed, by Theorem 16.18, $\partial U \subset \partial(\mathbb{R}^2\backslash\varphi^*) = \partial(\varphi^*) \subset \varphi^*$.

(iv) Take $R > 0$ such that $\varphi^* \subset \{x \in \mathbb{R}^2 : \|x\| \leq R\}$. The set $\{x \in \mathbb{R}^2 : \|x\| > R\}$ is path connected [Exercise 16.H(a)], hence contained in one component of $\mathbb{R}^2\backslash\varphi^*$. Thus, $\mathbb{R}^2\backslash\varphi^*$ has exactly one unbounded component.

(v) Let T be a homeomorphism $\mathbb{R}^2 \to \mathbb{R}^2$. Then, $T \circ \varphi$ is a curve in $\mathbb{R}^2$ and $(T \circ \varphi)^* = T(\varphi^*)$. If U is a component of $\mathbb{R}^2\backslash\varphi^*$, then $T(U)$ is a component of $\mathbb{R}^2\backslash(T \circ \varphi)^*$ and $\partial\big(T(U)\big) = T(\partial U)$. These observations will allow us to impose additional conditions on φ without jeopardizing the generality. Thus, when we find it convenient to assume $0 \in \varphi^*$ or $\varphi^* \subset \{x : \|x\| < 1\}$, we may do so.

(vi) Similarly, when it seems to be useful, we may assume the domain of our curve φ to be the interval $[0, 1]$.

Theorem 16.21

If φ is a Jordan curve in $\mathbb{R}^2$ that is not closed, then $\mathbb{R}^2\backslash\varphi^*$ is connected.

Proof

Let $\varphi : [0, 1] \to \mathbb{R}$ be a Jordan curve; let U be the unbounded component of $\mathbb{R}^2\backslash\varphi^*$ [16.20(iv)]. We propose to prove that U is all of $\mathbb{R}^2\backslash\varphi^*$; then, $\mathbb{R}^2\backslash\varphi^*$ will be connected. Thus, we are done if we can prove that every point of $\mathbb{R}^2$ lies in $\varphi^* \cup U$. One easily sees that it suffices to show that $0 \in \varphi^* \cup U$ [an application of 16.20(v)].

Furthermore, as φ^* is a bounded set, we may assume

$$\varphi^* \subset \{x \in \mathbb{R}^2 : \|x\| < 1\}. \tag{1}$$

By Corollary 13.23, $\varphi(t) \mapsto t$ is a continuous function $\varphi^* \to [0, 1]$, that can be extended to a continuous function $f : \mathbb{R}^2 \to [0, 1]$ [Tietze's Extension Theorem, (15.15)]. Setting $F := \varphi \circ f$, we obtain

$$F : \mathbb{R}^2 \to \varphi^* \text{ is continuous,}$$

$$F(x) = x \text{ for } x \in \varphi^*.$$

$\mathbb{R}^2$ is the union of two closed sets, $\mathbb{R}^2 \backslash U$ and U^-, the complement and the closure of U, respectively. For x in their intersection, we obtain $x \in \partial U \subset \varphi^*$ [16.20(iii)] and, consequently, $F(x) = x$. Thus, we can define $G : \mathbb{R}^2 \to \mathbb{R}^2$ by

$$G(x) := \begin{bmatrix} F(x) & \text{if } x \in \mathbb{R}^2 \backslash U, \\ x & \text{if } x \in U^-. \end{bmatrix}$$

According to the Glue Lemma (12.23), G is continuous. Observe that

$$G(x) \in \varphi^* \cup U \text{ for all } x \in \mathbb{R}^2. \tag{2}$$

Indeed, if $x \in \mathbb{R}^2 \backslash U$, then $G(x) = F(x) \in \varphi^*$; if $x \in U$, then $G(x) = x \in U$.

For $x \in \mathbb{R}^2$ with $\|x\| = 1$, we have $x \in U$ [see (1), above], so that $G(x) = x$. It follows from Exercise 4.E(ii) that 0 is a value of G. Thus,

$$0 \in \varphi^* \cup U$$

which is what we wished to prove.   ∎

Corollary 16.22
Let U be a nonempty open subset of $\mathbb{R}^2$ and suppose there is a Jordan curve φ with $\partial U \subset \varphi^*$. Then, U is unbounded and $U = \mathbb{R}^2 \backslash \partial U$.

Proof
The interior of φ^* is empty (Exercise 16.I). In particular, $U \not\subset \varphi^*$, Choose $a \in U$, $a \notin \varphi^*$. For every $x \in \mathbb{R}^2 \backslash \varphi^*$, by the theorem there is a curve in $\mathbb{R}^2 \backslash \varphi^*$ connecting a and x. Such a curve does not meet ∂U, so $x \in U$ (Lemma 16.8).

We see that $\mathbb{R}^2 \backslash \varphi^* \subset U$. It follows that U is unbounded. Since φ^* has empty interior, it also follows that the closure of U is $\mathbb{R}^2$. Hence, $U \cup \partial U = \mathbb{R}^2$ and, as $U \cap \partial U = \varnothing$, $U = \mathbb{R}^2 \backslash \partial U$.   ∎

From Corollary 16.22 we obtain part of the Jordan Theorem:

Corollary 16.23
Let ω be a closed Jordan curve in $\mathbb{R}^2$. If U is a component of $\mathbb{R}^2 \backslash \omega^*$, then $\partial U = \omega^*$.

Proof
It follows from Theorem 16.18 that $\partial U \subset \partial(\mathbb{R}^2 \backslash \omega^*) = \partial \omega^* \subset \omega^*$. Now *suppose $\omega^* \not\subset \partial U$.* For simplicity, assume the domain of ω is the interval

[0, 1]. As ∂U is closed, the set

$$\{t \in [0, 1] : \omega(t) \in \partial U\}$$

is a closed subset of $[0, 1]$, not equal to $[0, 1]$. Hence, there exists a subinterval $[r, s]$ of $[0, 1]$ such that

$$r < t < s \implies \omega(t) \notin \partial U$$

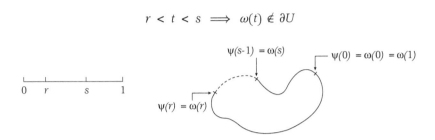

Define $\psi : [s-1, r] \to \mathbb{R}^2$ by

$$\psi(t) := \begin{bmatrix} \omega(t+1) & \text{if} & s-1 \leq t \leq 0, \\ \omega(t) & \text{if} & 0 \leq t \leq r. \end{bmatrix}$$

Then, ψ is a Jordan curve and $\partial U \subset \psi^*$. By Corollary 16.22, $\partial U = U^c \supset \omega^*$. *Contradiction.* ∎

16.24
Our proof of the Jordan Theorem is based on an extension of Theorem 4.26 to an "infinite rectangle."

Let A be the strip $[-1, 1] \times \mathbb{R}$ in $\mathbb{R}^2$; let φ be a Jordan curve in A, connecting the points $(-1, 0)$ and $(1, 0)$, such that all points of φ^* except these end points lie in the open set $(-1, 1) \times \mathbb{R}$.

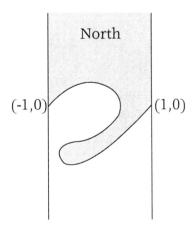

All points of A whose second coordinates are sufficiently large belong to the same path component of $A\backslash\varphi^*$. We denote this path component by

$$(A\backslash\varphi^*)_N;$$

its points are said to lie north of φ^*. In a similar way, we define the southern path component

$$(A\backslash\varphi^*)_S$$

whose points lie South of φ^*.

Theorem 16.25
In the situation described above, $A\backslash\varphi^*$ has precisely two path components: $(A\backslash\varphi^*)_N$ *and* $(A\backslash\varphi^*)_S$.

Proof
We have to show that $(A\backslash\varphi^*)_N$ and $(A\backslash\varphi^*)_S$ are distinct and that their union is all of $A\backslash\varphi^*$.

The first follows easily from Theorem 4.26. Indeed, *suppose* $(A\backslash\varphi^*)_N = (A\backslash\varphi^*)_S$. Take $b \in (0, \infty)$ so large that $\varphi^* \subset [-1, 1]\times[-b, b]$. Then, $(0, b) \in (A\backslash\varphi^*)_N$ and $(0, -b) \in (A\backslash\varphi^*)_S$, so there is a curve γ in $A\backslash\varphi^*$ connecting $(0, b)$ and $(0, -b)$. For some number $b' > b$, we have $\gamma^* \subset [-1, 1] \times [-b', b']$.

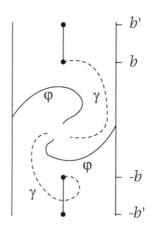

Then, the points $(0, b')$ and $(0, -b')$ can be connected by a curve in $A\backslash\varphi^*$ that stays within the rectangle $[-1, 1] \times [-b', b']$. This *contradicts* Theorem 4.26.

Next, take $p \in A\backslash\varphi^*$; we show that it lies either north or south of φ^*. As $\mathbb{R}^2\backslash\varphi^*$ is (path) connected (Theorem 16.20), there is a curve λ in $\mathbb{R}^2\backslash\varphi^*$ connecting p with a point outside A. By Lemma 16.8 there is a curve in $\lambda^* \cap A$, hence in $A\backslash\varphi^*$, connecting p with a point q of ∂A. Now, q is neither

$(-1, 0)$ nor $(1, 0)$ (since $q \in \lambda^* \subset \mathbb{R}^2 \backslash \varphi^*$), so q lies north or south of φ^*. Then, so does p. ∎

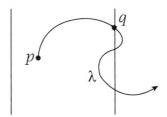

16.26

If $\omega : [0, 1] \to \mathbb{R}^2$ is a closed Jordan curve, then the formulas

$$\varphi(t) := \omega(\tfrac{t}{2}), \quad \psi(t) := \omega(1 - \tfrac{t}{2}) \quad (0, \leq t \leq 1)$$

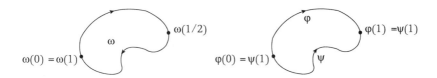

define two Jordan curves, φ and ψ. They have the same end points; apart from these, φ^* and ψ^* are disjoint, and $\omega^* = \varphi^* \cup \psi^*$.

Conversely, if we start with two Jordan curves φ and ψ, such that φ and ψ have the same end points but otherwise φ^* and ψ^* are disjoint, then we can find a closed Jordan curve ω with $\omega^* = \varphi^* \cup \psi^*$. (Namely, ...?)

Thus, for certain purposes, a couple of Jordan curves, suitably linked, will serve as well as one closed Jordan curve. This idea will be applied in 16.27. There, we will consider two linked Jordan curves, φ and ψ, and see how they divide the plane. The results will be used in 16.28.

For the time being, we restrict ourselves to the special situation described in (∗) and (∗∗), below. Occasionally, our language will be a bit informal, but there is not going to be any ambiguity or imprecision.

A is the vertical strip $[-1, 1] \times \mathbb{R}$.

e is the point $(1, 0)$.

φ and ψ are Jordan curves, both connecting $-e$ and e.   (∗)

$-e$ and e are the only points in $\varphi^* \cap \psi^*$; except for them,
    φ^* and ψ^* are contained in the open set $(-1, 1) \times (-1, 1)$.

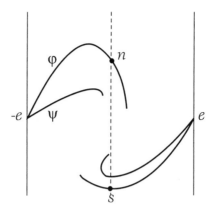

The set $(y\text{-axis}) \cap (\varphi^* \cup \psi^*)$ is bounded, closed, and nonempty; it follows easily that it has a highest point, n, and a lowest point, s. We assume

$$n \in \varphi^*. \tag{**}$$

(The alternative assumption, $n \in \psi^*$, is, of course, not really different.)

16.27

(I) Walking to the north from n, one arrives in the region $(A \setminus \psi^*)_N$ without touching upon ψ^*; hence, $n \in (A \setminus \psi^*)_N$. Every point of φ^* except $-e$ and e can be reached from n by a walk along φ^* that does not cross ψ^* either, so

$$\varphi^* \setminus \{-e, e\} \text{ lies north of } \psi^*. \tag{1}$$

Similarly, *if* $s \in \varphi^*$, then $\varphi^* \setminus \{-e, e\}$ lies south of ψ^*, which is impossible in view of (1) and Theorem 16.25. Thus, $s \notin \varphi^*$ so that $s \in \psi^*$. From this, it follows that

$$\psi^* \setminus \{-e, e\} \text{ lies south of } \varphi^* \tag{2}$$

precisely as (1) follows from (**).

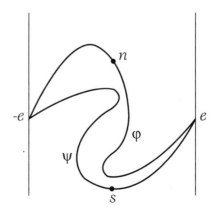

(II) Our next step is to show that $\mathbb{R}^2\backslash(\varphi^* \cup \psi^*)$ *has a bounded component.*
      The sets (y-axis) $\cap \varphi^*$ and (y-axis) $\cap \psi^*$ are closed, bounded, nonempty, and mutually disjoint. By an elementary consideration, there exist points a and b with

$$a \in (\text{y-axis}) \cap \varphi^*, \quad b \in (\text{y-axis}) \cap \psi^*,$$

between a and b there is no point of $\varphi^* \cup \psi^*.$ (3)

Choose a point p between a and b.

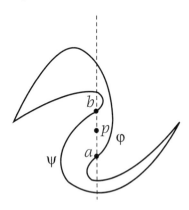

Then, $p \in \mathbb{R}^2\backslash(\varphi^* \cup \psi^*)$. Let U_0 be the component of $\mathbb{R}^2\backslash(\varphi^* \cup \psi^*)$ that contains p. We wish to show that U_0 is bounded.

Suppose U_0 is unbounded. Then U_0 is *the* unbounded component of $\mathbb{R}^2\backslash(\varphi^* \cup \psi^*)$, from which $U_0 \not\subset A$. There is a curve in U_0 connecting p with a point outside A. Hence, by Lemma 16.8, there is a curve λ in $U_0 \cap A$, hence in $(A\backslash\varphi^*) \cap (A\backslash\psi^*)$, that connects p with a point q on one of the vertical lines bounding A.

From a we walk due north or south to p, then follow λ^* until we reach q. Our walk has not touched upon ψ^* and has started north of ψ^*. [This follows from (1).] Thus, q lies north of ψ^* and its second coordinate must be positive. Similarly, we can go from b to p and then to q without meeting φ^*; as b lies south of φ^*, so must q and its second coordinate is negative. *Contradiction.*

(III) Now we can prove that $\mathbb{R}^2\backslash(\varphi^* \cup \psi^*)$ *has precisely one bounded component.*
      Take a, b, p, U_0 as in (II). We know that U_0 is a bounded component of $\mathbb{R}^2\backslash(\varphi^* \cup \psi^*)$. *Suppose* there is another bounded component, U. Then, $U \cap U_0 = \varnothing$.

Recall from (*) that $\varphi^* \cup \psi^*$ is contained in the "square" $[-1, 1] \times (-1, 1)$. Let f be the point $(0, 1)$ of $\mathbb{R}^2$. This f lies north of φ^* and ψ^*, whereas $-f$ lies south of them. The walk

      from f to n via the y-axis,
      from n to a via φ^*,
      from a to b via the y-axis,

from b to s via ψ^*,
from s to $-f$ via the y-axis

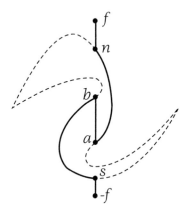

yields a curve γ connecting f with $-f$. The points due north of n and those due south of s lie in the *unbounded* component of $\mathbb{R}^2 \backslash (\varphi^* \cup \psi^*)$ and therefore not in U. The points of φ^* and ψ^* do not lie in U either; and the points between a and b lie in the set U_0 that is disjoint from U. Thus,

$$\gamma^* \cap U = \varnothing.$$

Let B be the horizontal strip $\mathbb{R} \times [-1, 1]$.

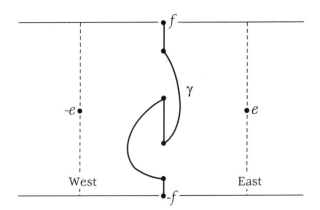

Proceeding as in Theorem 16.25, we see that $B \backslash \gamma^*$ has two path components, $(B \backslash \gamma^*)_E$ and $(B \backslash \gamma^*)_W$, containing $-e$ and e, respectively. As γ^* is a closed set, all points near $-e$ lie in $(B \backslash \gamma^*)_E$, all points near e in $(B \backslash \gamma^*)_W$.

Now it is easy to obtain a closed Jordan curve ω with $\omega^* = \varphi^* \cup \psi^*$. (See 16.26.) It follows from Corollary 16.23 that $\varphi^* \cup \psi^* = \omega^* = \partial U$.

In particular, $-e$ and e lie in ∂U, hence in U^-. Choose x and y in U so close to $-e$ and e that $x \in (B \backslash \gamma^*)_E$ and $y \in (B \backslash \gamma^*)_W$. Then, x and y can be connected by a curve in U but not by any curve in $B \backslash \gamma^*$. Because $U \subset B$, we see that

$$\gamma^* \cap U \neq \varnothing.$$

Contradiction.

16.28

It is now easy to prove the Jordan Closed Curve Theorem (16.19). Let ω be a closed Jordan curve in $\mathbb{R}^2$.

(I) First, assume

$$\omega^* \subset [-1, 1] \times (-1, 1), \qquad \omega^* \cap \partial A = \{-e, e\}$$

with $A, e, -e$ as in 16.26. In view of considerations such as set out in 16.20, we may assume the domain of ω to be the interval $[0, 1]$, and we may also assume $\omega(0) = -e$, $\omega(\frac{1}{2}) = e$. The formulas

$$\varphi(t) := \omega(\tfrac{t}{2}), \quad \psi(t) := \omega(1 - \tfrac{t}{2}) \qquad (0 \leq t \leq 1)$$

define Jordan curves φ and ψ satisfying $(*)$ of 16.26 with $\omega^* = \varphi^* \cup \psi^*$. By, if necessary, interchanging the names " φ " and " ψ ," we can ensure $(**)$, too. Then, the conclusion of the Jordan Theorem is guaranteed by Corollary 16.23 and by (II) and (III) of 16.27.

(II) Now we turn to the general case. We are done if we can find a homeomorphism $T : \mathbb{R}^2 \to \mathbb{R}^2$ such that

$$T(\omega^*) \subset [-1, 1] \times (-1, 1), \quad T(\omega^*) \cap \partial A = \{-e, e\}.$$

[See 16.20(v).]

ω^* being a compact subset of $\mathbb{R}^2$, there exist a and b in ω^* with

$$\|a - b\| \geq \|x - y\| \quad \text{for all } x, y \in \omega^*.$$

There is a map $T_1 : \mathbb{R}^2 \to \mathbb{R}^2$ that is a composition of a translation, a rotation, and a scalar multiplication such that

$$T_1(a) = -e, \quad T_1(b) = e.$$

(In terms of complex numbers, T_1 is the map $z \mapsto \frac{2z - (a+b)}{b-a}$.) This T_1 multiplies all distances with the same factor, so

$$2 = \|(-e) - e\| \geq \|u - v\| \quad \text{for all } u, v \in T_1(\omega^*).$$

In particular, for all $u \in T_1(\omega^*)$ we have $\|u - e\| \leq 2$ and $\|u + e\| \leq 2$. It follows that $T_1(\omega^*)$ is contained in the area shaded in this picture:

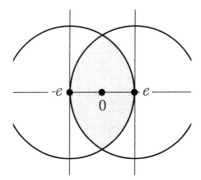

If T_2 is the linear map with matrix

$$\begin{pmatrix} 1 & 0 \\ 0 & \frac{1}{2} \end{pmatrix}$$

(T_2 shrinks in the vertical direction), then $T := T_2 \circ T_1$ satisfies the
requirements. ∎

Extra: Continuous Deformation of Curves

A problem one is often faced with in Topology is to determine whether two
given topological spaces are homeomorphic. The obvious way to prove
that they are is by constructing a homeomorphism, as we did in Example
4-19. To show that they are not, one usually looks for a "topological" property that one of the spaces has and the other does not have. For instance,
$[0, 1]$ is not homeomorphic to $\mathbb{R}$ since $[0, 1]$ is compact, whereas $\mathbb{R}$ is not.
$[0, 1]$ is not homeomorphic to the circle Γ because removal of a suitable
point (e.g., $\frac{1}{2}$) from $[0, 1]$ creates a disconnected space, whereas removal
of any one point from Γ does not.

Often, more sophisticated methods are required. Is the plane, $\mathbb{R}^2$,
homeomorphic to the "punctured plane," $\mathbb{R}^2 \backslash \{0\}$? To decide that we need
a construction that we have already used in 4.21, our informal proof of
Brouwer's Fixed Point Theorem. There, for every number r in $[0, 1]$ we had
a closed curve $\beta_r : [0, 1] \to \mathbb{R}^2$ and by varying r, we let the curve change
gradually. Such a family $(\beta_r)_{r \in [0,1]}$ of closed curves is called a "continuous
deformation of the curve β_0 into the curve β_1."

The general definition runs as follows. Suppose X is a topological space
and α and β are closed curves $[0, 1] \to X$. A *continuous deformation of α
into β* (or a *homotopy from α to β*) is a continuous map

$$(r, s) \mapsto \gamma_r(s)$$

of $[0, 1] \times [0, 1]$ into X such that

$$\gamma_0 = \alpha, \quad \gamma_1 = \beta,$$

$$\gamma_r(0) = \gamma_r(1) \text{ for every } r.$$

Then, for each r, γ_r is a closed curve in X. As r runs from 0 to 1, the curve α is gradually distorted until β is obtained.

α is called *homotopic* to β if such a continuous deformation exists. Then, β is also homotopic to α, so that one may reasonably say that α and β "are homotopic."

In $\mathbb{R}^2$, every closed curve (with domain $[0, 1]$) is homotopic to every other one. Indeed, if α and β are closed curves $[0, 1] \to \mathbb{R}^2$, the formula

$$\gamma_r(s) = r\alpha(s) + (1-r)\beta(s) \quad (s, t \in [0, 1])$$

defines a homotopy from β to α.

In the topological space $\mathbb{R}^2 \backslash \{0\}$, the situation is more interesting. It follows directly from Lemma 4.23 that *any two homotopic curves in $\mathbb{R}^2 \backslash \{0\}$ have the same winding number.* (It is to be understood that the connecting homotopy takes all of its values in $\mathbb{R}^2 \backslash \{0\}$.) Thus, the curves

$$s \mapsto (\cos 2\pi s, \sin 2\pi s) \quad (s \in [0, 1])$$

and

$$s \mapsto (1, 0) \quad (s \in [0, 1])$$

are not homotopic in $\mathbb{R}^2 \backslash \{0\}$.

It follows from the above that $\mathbb{R}^2$ and $\mathbb{R}^2 \backslash \{0\}$ are not homeomorphic.

The same technique can be used to prove, for instance, that the unit sphere in $\mathbb{R}^3$ is not homeomorphic to a torus. It is not too difficult to show that on the sphere, every closed curve is homotopic to a point curve, whereas on the torus, continuous deformation of a curve that "goes around" can only produce curves with the same property.

The concept of homotopy has led to various generalizations and theories, implying, e.g., that for $n \geq 2$, the spaces $\mathbb{R}^n$ and $\mathbb{R}^n \backslash \{0\}$ are not homeomorphic. But here we are at the beginnings of Algebraic Topology.

Further Reading

Rotman, Joseph J., *An Introduction to Algebraic Topology*, Springer-Verlag, New York, 1988.

Greenberg, M. and J.R. Harper, *Algebraic Topology: A First Course*, Benjamin Cummings, Reading (Mass.), 1981.

Exercises

16.A. Let A be a subset of a topological space X. Prove:

$$\mathbf{1}_A \text{ is continuous} \iff \partial A = \varnothing \iff A \text{ is closed and open.}$$

16.B. (i) Let Γ be the circle $\{x \in \mathbb{R}^2 : \|x\| = 1\}$. Show that for every $a \in \Gamma$ the set $\Gamma \setminus \{a\}$ is connected. Deduce that Γ and $[0, 1]$ are not homeomorphic.
   (ii) Prove (in a similar way?) that the intervals $(0, 1)$ and $(0, 1]$ are not homeomorphic.

16.C. Let A and B be nonempty closed subsets of a topological space X such that $A \cup B$ and $A \cap B$ are connected. Show that A and B must be connected, too. (Hint: First remark that $A \cap B \neq \varnothing$; then argue by contradiction.)
       What if A and B are not assumed to be closed?

16.D. Prove that if (X,ω) is a connected topological space and if ω' is a topology on X that is weaker than ω, then (X,ω') is connected.

16.E. (i) Let A be a connected subset of $[0, 1]$ containing more than one element. Let $a := \inf A$, $b := \sup A$. Show that $(a, b) \subset A \subset [a, b]$.
   (ii) Describe the connected subsets of $\mathbb{R}$.

16.F. (i) Show that if a connected metric space has more than one point, then it has uncountably many points. (Use the Tietze-Urysohn Extension Theorem.)
   (ii) Show that the topological space described in Exercise 15.H, which is Hausdorff and countably infinite, is connected.

16.G. (Compare Lemma 16.8.) Let C be a connected subset of a topological space X. Suppose Y is a subset of X such that $C \cap \partial Y = \varnothing$. Show that Y is contained either in the interior of Y or in the interior of $X \setminus Y$.

16.H. Which of the following subsets of $\mathbb{R}^2$ are path connected?
   (a) $\{x \in \mathbb{R}^2 : \|x\| > 37\}$.
   (b) $\{(s, t) \in \mathbb{R}^2 : t \leq s^2\}$.
   (c) $\{(e^{-t} \cos t, e^{-t} \sin t : t > 0\} \cup \{(0, 0)\}$. (Sketch! Note that if J is an interval and $\varphi : J \to \mathbb{R}^2$ is continuous, then $\varphi(J)$ is path connected.)

16.I. Let $\varphi : [0, 1] \to \mathbb{R}^2$ be continuous and injective, $\varphi^* := \{\varphi(t) : t \in [0, 1]\}$.
   (i) For $a \in \mathbb{R}^2$ and $\varepsilon > 0$, put $\Gamma_\varepsilon(a) := \{x \in \mathbb{R}^2 : \|x-a\| = \varepsilon\}$. Show that there can exist no a and ε such that $\Gamma_\varepsilon(a) \subset \varphi^*$. (Observe that by Corollary 13.23, φ is a homeomorphism of $[0, 1]$ onto φ^*.)
   (ii) Show that, as a subset of $\mathbb{R}^2$, φ^* has empty interior.

16.J. Show that if X and Y are connected topological spaces, then $X \times Y$ is connected.

16.K. We use the word "clopen" as an abbreviation for "closed and open." Let a be a point of a compact Hausdorff space X. Show that the intersection of all clopen subsets of X that contain a is precisely the component of a. [Hint: Let A be that intersection. The component of a is connected; infer that it must be contained in A. Show that we are done if we can prove A to be connected. Suppose it is not. As A is closed, it is a union of two disjoint nonempty closed subsets of X, A_1 and A_2. By normality (Theorem 15.7), there exist disjoint open set U_1 and U_2 with $A_1 \subset U_1, A_2 \subset U_2$. Use compactnss to show that $U_1 \cup U_2$ must contain a clopen set U containing a. Prove $U \cap U_1$ and $U \cap U_2$ to be clopen and derive a contradiction.]

16.L. Let X be a topological space. The conditions "X has only one component" and "X has only one closed-and-open subset" are equivalent. This observation, in conjunction with Theorem 16.12, suggests that components are always open. Give an example of a topological space with a component that is not open.

16.M. In $\mathbb{R}^2$, let $e := (1, 0)$, $C_n := \{(x, n^{-1}) : -1 \le x \le 1\}$ and

$$X := \bigcup_{n \in \mathbb{N}} C_n \cup \{e\} \cup \{-e\}.$$

Make a sketch of the set X. Show that $f(-e) = f(e)$ for every continuous $f : X \to \{0, 1\}$ and that, nevertheless, $-e$ and e lie in distinct components of X.

17

CHAPTER

Tychonoff's Theorem

17.1

We have had occasion before to use the product $X_1 \times X_2$ of two sets, X_1 and X_2. This $X_1 \times X_2$ is defined to be the set of all pairs (x_1, x_2) with $x_1 \in X_1$ and $x_2 \in X_2$. In a similar way, one makes the product of three or more sets: If $X_1, \ldots, X_n$ are sets, then $X_1 \times \cdots \times X_n$ consists of all n-tuples $(x_1, \cdots, x_n)$ with $x_i \in X_i$ for each i. Even for an infinite sequence $X_1, X_2, \ldots$ of sets, it makes sense to introduce the product $X_1 \times X_2 \times \cdots$, or

$$\prod_{i \in \mathbb{N}} X_i$$

as the set of all sequences $(x_1, x_2, \ldots)$ with $x_i \in X_i$ for all i.

For example, if A is a set, then $\prod_{i \in \mathbb{N}} A$ will simply be the set of all sequences of elements of A. In 1.9, we have considered such a sequence as a map $\mathbb{N} \to A$. In this language, $\prod_{i \in \mathbb{N}} A$ is the set of all such maps. Furthermore, if $X_1, X_2, \ldots$ are subsets of A, then $\prod_{i \in \mathbb{N}} X_i$ consists of all elements of $\prod_{i \in \mathbb{N}} A$ whose ith "coordinates" lie in X_i ($i \in \mathbb{N}$), so we may regard $\prod_{i \in \mathbb{N}} X_i$ as the set of all maps $f : \mathbb{N} \to A$ with $f(1) \in X_1, f(2) \in X_2, \ldots$, the map f being identified with the sequence $(f(1), f(2), \ldots)$.

17.2

It turns out to be useful to define products of even larger families of sets. Let A and I be sets. By

$$\prod_{i \in I} A \quad \text{or} \quad A^I,$$

we denote the set of all maps $I \to A$. Furthermore, if $(X_i)_{i \in I}$ is a family of subsets of A, we define

$$\prod_{i \in I} X_i := \{f \in A^I : f(i) \in X_i \text{ for each } i \in I\}. \qquad (*)$$

This set is the *Cartesian product* of $(X_i)_{i \in I}$.

Observe the (trivial) fact that $\prod_{i \in I} X_i$ depends only on I and the family $(X_i)_{i \in I}$, not on A. Indeed, if B is another set containing every X_i, then

$$\{f \in A^I : f(i) \in X_i \text{ for every } i \in I\}$$

is identical with

$$\{f \in B^I : f(i) \in X_i \text{ for every } i \in I\}.$$

This means that we have actually defined $\prod_{i \in I} X_i$ for *any* family of sets $(X_i)_{i \in I}$: All we need is one set A containing all X_i's as subsets, and for that we can always take their union. Explicitly, for any family $(X_i)_{i \in I}$ of sets, $\prod_{i \in I} X_i$ is the set of all maps $f : I \to \bigcup_{i \in I} X_i$ that have the property

$$f(i) \in X_i \text{ for every } i \in I.$$

17.3

Let $(X_i)_{i \in I}$ be a family of topological spaces, $X = \prod_{i \in I} X_i$. For every $i \in I$, we have a "coordinate map" $p_i : X \to X_i$:

$$p_i(f) := f(i) \quad (f \in X).$$

The *product topology* on X is the weak topology induced by the family $(X_i, p_i)_{i \in I}$. (See 14.3.) It is the weakest topology on X that makes all coordinate maps continuous.

17.4

(i) As we have seen in Theorem 14.5, a net $(f_\tau)_{\tau \in T}$ in X converges in the sense of the product topology to an element f of X if and only if $p_i(f_\tau) \to p_i(f)$ for every i. In other words,

$$f_\tau \to f \iff f_\tau(i) \to f(i) \text{ for every } i.$$

Convergence in the product topology is coordinatewise convergence. (Compare Example 12.15.)

A subbase for the product topology is

$$\{p_i^{-1}(U) : i \in I; U \subset X_i \text{ open }\}.$$

(See Exercise 17.B.)

(ii) If $I = \{1, 2\}$, then $X = X_1 \times X_2$; the product topology defined above is precisely the one introduced in 14.7.

(iii) If $X_i = \mathbb{R}$ for every i, then X is the set of all functions on I; the product topology is the topology of pointwise convergence. See [Example 14.6(ii)].

(iv) For each i, let Y_i be a subset of X_i, considered as a topological space under the relative topology. Then $\prod_{i \in I} Y_i$ is a subset of $\prod_{i \in I} X_i$. The product topology of $\prod_{i \in I} Y_i$ is precisely the relative topology induced by the product topology of $\prod_{i \in I} X_i$.

The product topology inherits certain properties from the topologies on the factor spaces. A famous result in this area is the following.

Theorem 17.5 (Tychonoff's Theorem)
Let $(X_i)_{i \in I}$ be a family of compact topological spaces. Then, $\prod_{i \in I} X_i$ is compact in the product topology.

A proof of this theorem requires an excursion into Set Theory: We need a result known as "Zorn's Lemma." If you are familiar with this lemma, you may now go to 17.12. Otherwise, you are in for some hard work.

17.6
Let S be a set. A collection γ of subsets of S is called a *chain* if

$$A, B \in \gamma \implies A \subset B \text{ or } B \subset A.$$

(The empty collection is taken to be a chain.) Every subcollection of a chain is a chain.

An example: The intervals $(-a, a)$ with $a > 0$ form a chain of subsets of $\mathbb{R}$. The collection of *all* intervals is no chain, since neither of the intervals $(0,2)$ and $(1,3)$ is contained in the other.

The set-theoretic tool we need is the following special case of *Zorn's Lemma*, to be proved in 17.11:

Lemma 17.7
Let S be a nonempty set. Let there be given a collection of subsets of S, called the "beautiful sets." Suppose

> *Every chain consisting of beautiful sets has a beautiful union.* $(*)$

Then there is a maximal beautiful set.

17.8
Some explanation is in order. A *maximal* beautiful set is a beautiful set M that is not contained in any strictly larger beautiful set, i.e., that has the property

> if A is beautiful and $M \subset A$, then $M = A$.

For an example to illustrate Condition $(*)$, let S be $\mathbb{R}^2$ and let us call a subset A of $\mathbb{R}^2$ "beautiful" if the function $x \longmapsto \|x\|$ is constant on A. Let

γ be a chain of beautiful sets; we prove $\bigcup \gamma$ to be beautiful. To this end, take $a, b \in \bigcup \gamma$; we wish to show that $\|a\| = \|b\|$. There exist $A \in \gamma$ and $B \in \gamma$ with $a \in A$ and $b \in B$. As γ is a chain, either $A \subset B$ or $B \subset A$; hence, either $a, b \in B$ or $a, b \in A$. In any case, $\|a\| = \|b\|$ because B and A are beautiful.

This proves (∗). The maximal beautiful sets whose existence the lemma guarantees are, of course, the circles centered at 0.

17.9
For subsets A and B of a set S, we put

$$A < B$$

if $A \subset B$ and $A \neq B$. If α is a collection of subsets of S and $C \subset S$, we put

$$\alpha_C := \{A \in \alpha : A < C\}.$$

Lemma 17.10
Let S be a set. For every collection γ of subsets of S, let there be given a subset $T(\gamma)$ of S. A nonempty collection α of subsets of S is called a "ladder" if it has the following two properties.

(a) *For every nonempty subcollection γ of α, we have $\bigcap \gamma \in \gamma$.*
(b) *For every $A \in \alpha$, we have $T(\alpha_A) = A$.*
  Define the collection ω by

$$A \in \omega \iff A \text{ belongs to some ladder.}$$

Then ω is a ladder.

  (The definition of "ladder" is entirely unnatural. Its only excuse is that it works. We use the word with due respect to the towers of Dugundji.)

Proof
First, observe that ladders exist: The one-set collection $\{T(\varnothing)\}$ is one. Condition (a) says that if γ is a nonempty subcollection of a ladder, then among the sets belonging to γ, there is a smallest one.

(I) Every ladder is a chain. Indeed, let α be a ladder and let $A, B \in \alpha$. Upon applying (a) to $\gamma := \{A, B\}$, we see that $A \cap B$ is either A or B, so either $A \subset B$ or $B \subset A$.
(II) If α is a ladder, $C \subset S$, and $\alpha \neq \alpha_C$, then $T(\alpha_C) \in \alpha$. Proof: Let γ be the nonempty collection $\{A \in \alpha : A \not< C\}$ and let B be the smallest set belonging to γ. Then, $B \in \gamma \sqsubset \alpha$, so we are done if $T(\alpha_C) = B$, which will be the case if $\alpha_C = \alpha_B$. Take $A \in \alpha$; we prove $A < C \iff A < B$. If $A < B$, then $B \not\subset A$, so $A \not\in \gamma$, from which $A < C$. On the other hand, $B \in \gamma$, so $B \not< C$; hence, if $A < C$, then $B \not\subset A$, so that $A < B$ according to (I).

(III) If α and β are ladders, then $\alpha \sqsubset \beta$ or $\beta \sqsubset \alpha$. Proof: *Suppose* not. The collections $\alpha \backslash \beta := \{A \in \alpha : A \notin \beta\}$ and $\beta \backslash \alpha := \{B \in \beta : B \notin \alpha\}$ are both nonempty. Let A be the smallest set in $\alpha \backslash \beta$, and B the smallest set in $\beta \backslash \alpha$. Obviously, $A \neq B$, so we may assume $A \not\subset B$.

If C is a set and $C < B$, then $A \not\subset C$ (since $A \not\subset B$) and $B \not\subset C$, so $C \notin \alpha \backslash \beta$ and $C \notin \beta \backslash \alpha$. Therefore, $\alpha_B = \beta_B$. As $\alpha \not\sqsubset \beta$, it follows that $\alpha_B \neq \alpha$. Hence, by (II), $T(\alpha_B) \in \alpha$. But $T(\alpha_B) = T(\beta_B) = B \notin \alpha$. *Contradiction.*

(IV) If α is a ladder and $C \subset S$, $C \in \alpha$, then $\alpha' := \{A \in \alpha : A \subset C\}$ is a ladder. (It is straightforward that $\alpha'_A = \alpha_A$ for every $A \in \alpha'$.)

(V) ω is a chain. Proof: Let $A, B \in \omega$. There exists ladders α and β with $A \in \alpha$ and $B \in \beta$. Because of (III), we may assume $\alpha \sqsubset \beta$. Then $A, B \in \beta$, so $A \subset B$ or $B \subset A$ by (II).

(VI) If α is a ladder and $A \in \alpha$, then $\omega_A = \alpha_A$. Proof: As $\alpha \sqsubset \omega$, trivially we have $\alpha_A \sqsubset \omega_A$. For the reverse, take B_0 belonging to ω_A. Choose a ladder β with $B_0 \in \beta$. By (IV), $\beta' := \{B \in \beta : B \subset B_0\}$ is a ladder. As $B_0 < A$, we see that $A \notin \beta'$, so $\alpha \not\sqsubset \beta'$. Then $\beta' \sqsubset \alpha$ because of (III), so that $B_0 \in \alpha$ and $B_0 \in \alpha_A$.

(VII) Now we can prove ω to be a ladder.

Let γ be a nonempty subcollection of ω. Take $A \in \gamma$ and let α be a ladder containing A. As ω is a chain, so is γ. If A is the smallest set belonging to γ, we are done. Otherwise, γ_A is a nonempty subcollection of ω_A, hence of α_A [by (VI)]. Then, γ_A contains a smallest set, which is also the smallest set in γ.

Finally, take $A \in \omega$. If α is a ladder containing A, then, again by (VI), $T(\omega_A) = T(\alpha_A) = A$.   ∎

17.11

With the aid of this lemma, we prove Lemma 17.7:

The union of the empty chain is $\varnothing$, so $\varnothing$ is a beautiful set.

Suppose there is no maximal beautiful set; i.e., for every beautiful set A, there exists a beautiful set B with $A < B$. To make Lemma 17.10 applicable, for every collection γ of subsets of S we choose a subset $T(\gamma)$ of S, subject only to these requirements:

$T(\gamma)$ is beautiful;

if $\bigcup \gamma$ is beautiful, then $\bigcup \gamma < T(\gamma)$.

Form the ladder ω as in Lemma 17.10. Let $O := T(\omega)$.

For every $A \in \omega$, we have $A = T(\omega_A)$, so A is beautiful. By the property (∗) of Lemma 17.7, $\bigcup \omega$ is beautiful, so $\bigcup \omega < O$. In particular, $O \notin \omega$.

However, by adding the set O to the collection ω, we obtain a collection $\omega' := \omega \cup \{O\}$ that is easily seen to be a ladder. [If $A \in \omega$, then $T(\omega'_A) = T(\omega_A) = A$, whereas $T(\omega'_O) = T(\omega) = O$.] Hence, by the definition of ω, we have $\omega' \sqsubset \omega$, so $O \in \omega$. *Contradiction.*   ∎

After this set-theoretic digression we return to Topology.

17.12
We proceed to prove Tychonoff's Theorem 17.5.

In this proof, we simply write "Π" instead of "$\prod_{i \in I}$."

Let the symbol "∞" represent a mathematical object that is no element of any X_i. For each i, put $X_i^\sim := X_i \cup \{\infty\}$. By adding the set $X_i^\sim$ to the given topology ω_i of X_i, we obtain a topology $\omega_i^\sim := \omega_i \sqcup \{X_i^\sim\}$ on $X_i^\sim$; then, ω_i is the relative topology on X_i induced by $\omega_i^\sim$. Consequently, the product topology on ΠX_i is the relative topology induced by the product topology on $\Pi X_i^\sim$.

In $X_i^\sim$, every net converges to ∞. Hence, in $\Pi X_i^\sim$ every net converges to the constant map $i \mapsto \infty$ $(i \in I)$.

Let $(f_\tau)_{\tau \in T}$ be a net in ΠX_i; we are done if we can prove it has a cluster point in ΠX_i. (See Theorem 13.13.) The set Z of all cluster points of the net in $\Pi X_i^\sim$ is nonempty; it contains, for instance, the constant map $i \mapsto \infty$ $(i \in I)$. We intend to prove

$$Z \cap \Pi X_i \neq \varnothing.$$

For $g_1, g_2 \in Z$, put $g_1 \succ g_2$ if

$$\text{for every } i \in I : \ g_2(i) = g_1(i) \text{ or } g_2(i) = \infty.$$

The relation $\succ$ is transitive. (See 10.6.) We call a subset A of Z "beautiful" if $\succ$ is inductive on A; i.e., if

$$g_1, g_2 \in A \implies \text{there is a } g \in A \text{ with } g \succ g_1, g \succ g_2.$$

If γ is a chain of beautiful sets, then $\bigcup \gamma$ is beautiful. (Proof: Let $g_1, g_2 \in \bigcup \gamma$. There exist $A_1, A_2 \sqsubseteq \gamma$ with $g_1 \in A_1$, $g_2 \in A_2$. We have $A_1 \subset A_2$ or $A_2 \subset A_1$; assume $A_1 \subset A_2$. Then $g_1, g_2 \in A_2$, so there is a g in A_2, hence in $\bigcup \gamma$, with $g \succ g_1, g \succ g_2$.) It follows from Lemma 17.7, our version of Zorn's Lemma, that there is a maximal beautiful set, $\overline{A}$, say.

If $i \in I$ and if $g_1, g_2 \in \overline{A}$, $g_1(i) \neq \infty$, and $g_2(i) \neq \infty$, then it follows from the inductivity of $\succ$ on $\overline{A}$ that $g_1(i) = g_2(i)$. Thus, there is a $\overline{g} \in \Pi X_i^\sim$ such that

$$\overline{g}(i) = g(i) \quad \text{whenever } i \in I, g \in \overline{A}, g(i) \neq \infty;$$

$$\overline{g}(i) = \infty \quad \text{whenever } i \in I \text{ and } g(i) = \infty \text{ for all } g \in \overline{A}.$$

We show that $\overline{g} \in Z \cap \Pi X_i$.

Under $\succ$, $\overline{A}$ is a directed set. It is essentially trivial that for each i, the net $(g(i))_{g \in \overline{A}}$ converges to $\overline{g}(i)$ in $X_i^\sim$; hence, the net $(g)_{g \in \overline{A}}$ converges to $\overline{g} \in \Pi X_i^\sim$. But Z, the set of all cluster points of $(f_\tau)_{\tau \in T}$, is closed in $\Pi X_i^\sim$. (See 13.11.) Therefore, $\overline{g} \in Z$.

To prove that $\overline{g} \in \Pi X_i$, take $j \in I$; we show $\overline{g}(j) \neq \infty$. Suppose $\overline{g}(j) = \infty$. The net $(f_\tau)_{\tau \in T}$ has a subnet converging (coordinatewise) to $\overline{g}$. By the compactness of X_j, this subnet in turn has a subnet $(f'_\sigma)_{\sigma \in S}$ such that $(f'_\sigma(j))_{\sigma \in S}$ converges to some x_j in X_j. Define $g' \in \Pi X_i^{\sim}$ by

$$g'(j) := x_j;$$

$$g'(i) := \overline{g}(i) \quad \text{for} \ i \neq j.$$

Then $f'_\sigma \to g'$ coordinatewise, so $g' \in Z$. For all $g \in \overline{A}$, we have $g' \succ \overline{g} \succ g$. It follows that $\overline{A} \cup \{g'\}$ is a beautiful set, hence is equal to $\overline{A}$. In particular, $g' \in \overline{A}$, so that (by the definition of $\overline{g}$) $\overline{g}(j) = g'(j) \neq \infty$. *Contradiction.* ∎

Example 17.13

A compact space with a sequence that has no convergent subsequence.

Let $\mathcal{P}(\mathbb{N})$ be the collection of all subsets of $\mathbb{N}$. By Tychonoff's Theorem, the space $X := \{0, 1\}^{\mathcal{P}(\mathbb{N})}$ is compact under the product topology. Define $f_1, f_2, f_3, \ldots \in X$ by

$$f_n(A) := \mathbf{1}_A(n) \qquad (n \in \mathbb{N}).$$

Suppose the sequence $(f_n)_{n \in \mathbb{N}}$ has a convergent subsequence $(f_{\alpha(n)})_{n \in \mathbb{N}}$. Then, for every $A \subset \mathbb{N}$ the number sequence $f_{\alpha(1)}(A), f_{\alpha(2)}(A), \ldots$, which is

$$\mathbf{1}_A(\alpha(1)), \mathbf{1}_A(\alpha(2)), \ldots,$$

converges. We get a contradiction by choosing $A := \{\alpha(2), \alpha(4), \alpha(6), \ldots\}$.

Extra: The Axiom of Choice

In Chapter 2, we have considered axioms for the real-number system, the rules of the game called Analysis. Playing the game, one uses not only those rules but also logic and set theory. Other branches of mathematics are played according to other rules, but the logic and the set theory remain the same. Both have been studied intensively and both are themselves games with fixed rules. The average mathematician does not bother with these any more than the average engineer is concerned with the axioms of Analysis, but that is deplorable. As a postscript to this chapter, we discuss one of the axioms of set theory.

Various axiom systems for set theory have been proposed. One of the most widely used is the system **ZF**, called after Ernst Zermelo and Adolf Fraenkel. The axioms of **ZF** basically are elementary ways to make new sets out of old. They will, for instance, guarantee that for any set X, the objects Y with $Y \subset X$ form again a set, which then is called $\mathcal{P}(X)$.

There is, however, one set-theoretic operation we have carried out repeatedly that is not validated by the rules of the **ZF**-game. It relies on the so-called Axiom of Choice. This axiom can be added to **ZF**, creating the game of set theory as it is usually played, but some mathematicians reject it because they find it makes the game meaningless.

Suppose A and B are nonempty sets. The fact that A is nonempty means that there exists an a with $a \in A$. Similarly, there exists a b with $b \in B$. Then the pair (a, b) is an element of the Cartesian product $A \times B$. Consequently, $A \times B$ is nonempty.

Similarly, Cartesian products of three, four, or nineteen nonempty sets are nonempty. But how about infinite products? Suppose $(A_i)_{i \in I}$ is any family of nonempty sets. The definition of the Cartesian product $A :=$ $\prod_{i \in I} A_i$ makes good sense within the framework of **ZF**, but is A nonempty? The intuitive approach is like this: For each i, there is an a_i in A_i; the map $i \mapsto a_i$ is an element of A. Actually, that is what we have done quite often in this book. One early instance is Lemma 2.16, where we proved that if $X \subset \mathbb{R}$, $c \in \mathbb{R}$, and c is adherent to X (i.e., $X \cap [c-\varepsilon, c+\varepsilon] \neq \varnothing$ for all $\varepsilon > 0$), then there is a sequence in X converging to c. Our reasoning was:

$$\text{For each } n \in \mathbb{N}, \text{ take a point } x_n \text{ in } X \cap \left[c - \frac{1}{n}, c + \frac{1}{n}\right].$$

Then $x_1, x_2, \ldots X$ and $x_n \to c$.

You see what happens: For each n, we know that the set $X_n := X \cap [c - \frac{1}{n}, c + \frac{1}{n}]$ is nonempty. What we need is a *sequence* $x_1, x_2, \ldots$ with $x_n \in X_n$ for each n, i.e., a function on $\mathbb{N}$ whose value at n lies in X_n (i.e., an element of $\prod_{n \in \mathbb{N}} X_n$).

The statement

If $(A_i)_{i \in I}$ is a family of nonempty sets,

then the Cartesian product $\prod_{i \in I} A_i$ is nonempty

is known as the *Axiom of Choice*,

AC.

Our proof of Lemma 2.16 relies on it.

But **AC** is not implied by the axioms of **ZF**. The trouble is that **AC** merely claims that a certain set is nonempty without providing a construction of an element of it. The **ZF** axioms are different in kind. For instance, they do not merely say that for every set X, there exists a set $\mathcal{P}(X)$ with certain properties; they precisely describe what objects are elements of $\mathcal{P}(X)$. In terms of **ZF**, you can completely describe the Cartesian product itself, but not any particular element of it. The axioms of **ZF** are *constructive*; the Axiom of Choice is much vaguer. It is so vague that a considerable number of mathematicians do not accept it. (Many others use it whenever it is convenient and simply refuse to think about it.)

That does not make **AC** worthless. In fact, one can show that, in a sense, it cannot do any harm.

Everyone who wants to do mathematics may choose his or her own axioms. However, it would be silly to take axioms that lead to contradictions. If from your axioms for certain objects a and b, you can prove both formulas "$a = b$" and "$a \neq b$" then you might not have made a mistake in your proofs you might have chosen an unfruitful axiom system. An axiom system that leads to contradictions is called *inconsistent*. For quite some time, there was a problem whether **ZF** together with **AC** might be inconsistent. In 1938, Kurt Gödel showed that it is not, provided that **ZF** by itself is not inconsistent. Thus, adding **AC** to **ZF** is not "wrong." (Somewhat disconcertingly, Paul Cohen proved in 1963 that if by non**AC** we mean "**AC** is false," then adding non**AC** to **ZF** does not create inconsistency either!)

In our extra to Chapter 2 we mentioned equivalence of axioms. Within the framework of our Axioms I-II-III-IV, Axiom V was equivalent to CT, the Connectedness Theorem. A similar situation occurs regarding the Axiom of Choice. There are various statements that, within **ZF**, are equivalent to **AC**. One of them is Tychonoff's Theorem, **T**. In 17.11, we proved it from **ZF** and **AC**. It is not particularly difficult to show that, conversely, **AC** follows from **ZF** and **T**. This is how it is done:

Let $(A_i)_{i \in I}$ be a family of nonempty sets. Choose an object ∞ that does not lie in any A_i. For each i, put $A_i^+ := A_i \cup \{\infty\}$. The collection of (three) sets

$$\{\varnothing, A_i^+, \{\infty\}\}$$

is a compact topology on A_i^+. By **T**, $A^+ := \prod_{i \in I} A_i^+$ is compact. For each i, put

$$X_i := \{f \in A^+ : f(i) \in A_i\}.$$

As A_i is closed in A_i^+, X_i is closed in A^+. Now,

$$\prod_{i \in I} A_i = \bigcap_{i \in I} X_i,$$

so that, by virtue of compactness, we are done if we can prove that the collection of sets $\{X_i : i \in I\}$ is finitely bound. Let F be a finite subset of I. Then, the finite product $\prod_{i \in F} A_i$ contains an element, f_0, say. Defining $f \in A^+$ by

$$f(i) := f_0(i) \quad \text{if } i \in F,$$
$$f(i) := \infty \quad \text{if } i \neq F,$$

we have $f \in X_i$ for each $i \in F$.

Let us agree to work with **ZF**.

From the above, we see that **AC** and **T** are equivalent. Observing that our proof of **T** in 17.11 ran via Lemma 17.7 (and Lemma 17.10, which does not use **AC**), we note that, actually, **AC** $\Longrightarrow$ Lemma 17.7 $\Longrightarrow$ **T** $\Longrightarrow$

AC, so that Lemma 17.7 and **AC** are equivalent, too. We have mentioned Zorn's Lemma as a more general statement than Lemma 17.7. For the benefit of readers who are used to work with partially ordered sets, we discuss Zorn's Lemma and show its equivalence with **AC**.

Zorn's Lemma, **Z**, is:

> If X is a nonempty partially ordered set in which
>
> every totally ordered subset has an upper bound,
>
> then X has a maximal element.

Lemma 17.7 is a special case: Simply let X be the collection of all beautiful sets, partially ordered by inclusion, and see what happens.

On the other hand, assuming 17.7, we can prove **Z**: Let S be the collection of all subsets of X. Say that a subset of X is "beautiful" if and only if it is totally ordered. It is not too hard to prove that $(*)$ of Lemma 17.7 is fulfilled. Hence, X has a maximal totally ordered subset, X_0. This X_0 has an upper bound, a, say. Then, a is easily seen to be a maximal element of X.

Thus, in **ZF**, *Zorn's Lemma is equivalent to the Axiom of Choice.*

How did people "invent" or "discover" **AC**? We gleaned most of the history of **AC** below from a very detailed answer to that question in the book by Moore, listed in Further Reading.

This history of the Axiom of Choice has many of its roots in the early stages of Analysis and Topology. The existence of mathematical objects of one generation is shaped by the knowledge of its ancestry as much as it molds the shape and form of things yet to come. Sometimes, in the course of new constructions, hidden arguments or assumptions creep in. Some of these hidden assumptions seem so plausible that they go undetected for many generations. Meanwhile, a network of consequences —let us call them unproved theorems—will result. At the moment that introspection into the assumptions of a previous age reveals the concealed weakness, the shock can be considerable. In mathematics, the shock was never as big as in 1904, when Zermelo revealed the Axiom of Choice.

It had intruded very slowly and well hidden, into arguments far before 1904. Since antiquity, it had been good practice (and no Axiom of Choice is needed) to choose one unspecified element from a set. In the early nineteenth century, mathematicians (e.g., Cauchy and Bolzano) started to choose an infinite number of unspecified elements from a set. By 1893, when Jordan published the second edition of his *Cours d'Analyse*, the usage of the hidden assumption was widespread.

At the same time, Cantor had moved his interests from Analysis to Set Theory. Initially, he proposed as a "law of thought" that every set can be well-ordered. (If you do not know what that means, sorry, we are not going into details here.Exercise 17.E provides the connection with **AC**.) His proposal found little approval and, around 1890, Cantor started to believe that he might be able to *prove* that law of thought. He was not successful. Then, in 1904, Zermelo provided a proof and he acknowledged right away that he was using the Axiom of Choice. The storm had been unleashed. The criticisms ranged from outright rejection (Baire and Borel in France, Brouwer in the Netherlands, and Peano and Levi in Italy) to words of caution (Lebesgue in France). The defense pointed to the abundant use of the Axiom before. Zermelo himself, in his 1904 paper, rightfully remarked that "it is used everywhere in mathematical deductions without hesitation." Others (Fréchet and Poincaré) were of the opinion that it was a mere matter of definition. Ironically, some of the harsher critics were only slow to recognize how they had (unknowingly) used the Axiom in their own work. Hadamard, who at the École Polytechnique had taken the torch from Jordan, was particularly adamant in his defense of the Axiom of Choice. Fréchet, a student of Hadamard's, published his thesis in 1906 on arbitrary sets equipped with a metric. Thus, the psychological barrier to study sets of which the elements need not be specified was taken at about the same time that arbitrary choices of unspecified elements was formally proposed as a valid enterprise. In the same period, Lebesgue revolutionized measure theory, again focusing on sets in general and axiomatics to streamline the results. On top of all that, Zermelo himself, in 1908, presented an axiomatic approach to Set Theory itself.

Hausdorff's 1914 treatise on Topology marks the end of an era. Between 1906 and 1908, Hausdorff became convinced by Zermelo's argument and he accepted the free use of the Axiom of Choice. It also was his opinion that one should not waste time on foundational discussions. Consequently, he did not keep book of when and how he used the Axiom. With that attitude, he set the tone of topological research to come. Topologists very quickly advanced with little or no qualms about the Axiom of Choice. In a period of less than 10 years, all sequential notions, as initiated by the study of metric spaces, were replaced by constructs in which sequential arguments were no longer fundamental. In 1923, Alexandroff and Urysohn proposed the definition of compactness which was about to replace Fréchet's notion of sequential compactness. In 1930, Tychonoff proved that every product of copies of $[0, 1]$ is compact, as a by-product of investigating the embedding of a Hausdorff space in a compact one. Čech saw in 1937 that these results could be generalized and proved what is now known as Tychonoff's Theorem. Kelley proved in 1950 that Tychonoff's Theorem is equivalent to the Axiom of Choice.

Further Reading

Moore, G. H., *Zermelo's Axiom of Choice, its Origins, Development and Influence*, Springer-Verlag, New York, 1982.

Exercises

17.A. Show that *the product of any family of Hausdorff spaces is Hausdorff.*

17.B. Let $(X_i)_{i \in I}$ be a family of topological spaces. By a *box* in $\prod_{i \in I} X_i$, we mean a set of the form $\prod_{i \in I} Y_i$ where Y_i is an open subset of X_i for each i. If I consists of two elements, these boxes form a base for the product topology (14.8.) It is easy to see that the same is true whenever the set I is finite. However:

Show that a box $\prod_{i \in I} Y_i$, if nonempty, is open in the product topology if and only if $Y_i \neq X_i$ for only finitely many i.

17.C. Let $(X_i)_{i \in I}$ be a family of topological spaces and $X = \prod_{i \in I} X_i$.
(i) Take $g \in X, j \in I$ and $x \in X_j$. Define $f : X_j \to X$ by

$$\big(f(x)\big)(i) := g(i) \quad \text{if } i \neq j,$$

$$\big(f(x)\big)(j) := x.$$

(Make a sketch for $I = \{1, 2\}$.) Show that f is continuous.
(ii) Take $g_0 \in X$ and let

$$G := \{g \in X : g(i) \neq g_0(i) \text{ for only finitely many } i\}.$$

Show that G is dense in X.
(iii) Now prove that *the product of any family of connected topological spaces is connected.*

17.D. Prove that *the product of countably many metrizable spaces is metrizable.* [By Example 10.3(i), products of uncountably many spaces give problems.]

Hint: For $n \in \mathbb{N}$ let X_n be a metrizable space. Put $X := \prod_{n \in \mathbb{N}} X_n$. We view an element x of X as a sequence $(x_1, x_2, \ldots)$. It follows from Exercise 5.I that for each n, there is a metric d_n on X_n that determines the given topology of X_n and has its values in $[0, 1]$. Define

$$D(x, y) := \sup\{n^{-1} d_n(x_n, y_n) : n \in \mathbb{N}\} \quad (x, y \in X).$$

Show that D is a metric on X.

In Exercise 12.C, we have put a metric d on $[0, 1]^{\mathbb{N}}$, determining the topology of pointwise convergence. Use this metric to prove that on X the D-topology is the product topology.

17.E. (An excursion into Set Theory, meaningful only if you have some experience with ordered sets.) An ordering of a set S is a *well-ordering* if every nonempty subset of S has a smallest element. Example: the usual ordering in $\mathbb{N}$.

If $(X_i)_{i \in I}$ is a family of nonempty sets and if a well-ordering of $\bigcup_{i \in I} X_i$ is given, then

$$i \mapsto \quad \text{the smallest element of } X_i$$

is an element of $\prod_{i \in I} X_i$, obtained without the Axiom of Choice.
The statement

Every nonempty set admits a well-ordering

is known as the Well-Ordering Theorem, **W**. By the above, **W** implies **AC** (assuming **ZF**). Actually,

$$\mathbf{W} \quad \Longleftrightarrow \quad \mathbf{AC}.$$

To see this, let S be a nonempty set. **AC** claims the existence of a map λ, assigning to every subset V of S that is not S itself an element $\lambda(V)$ of $S \backslash V$. For any nonempty collection γ of subsets of S, let $T(\gamma)$ be the subset of S given by

$$T(\gamma) := \bigcup \gamma \cup \{\lambda(\bigcup \gamma)\} \quad \text{if } \bigcup \gamma \neq S;$$

$$T(\gamma) := S \qquad\qquad\qquad \text{if } \bigcup \gamma = S.$$

Now proceed using the language of Lemma 17.10:

(i) Show: if α is a ladder, then so is $\alpha \cup \{T(\alpha)\}$.

(ii) Infer that $\bigcup \omega = S$.

(iii) Hence, for every $s \in S$, we can define $V(s)$ to be the smallest set in ω that contains s. Then $\bigcup \omega_{V(s)}$ does not contain s. Deduce that

$$T(\omega_{V(s)}) = \bigcup \omega_{V(s)} \cup \{s\} \qquad (s \in S).$$

(iv) Prove that $s \mapsto V(s)$ is an injection $S \to \omega$ and that the formula

$$s_1 \leq s_2 \text{ if and only if } V(s_1) \subset V(s_2)$$

defines a well-ordering $\leq$ of S.

IV

PART

Postscript

18
CHAPTER

A Smorgasbord for Further Study

18.1

Our theory is essentially completed. The present chapter is devoted to secondary matters.

First, we have often made use of nets. Indeed, in Chapter 11, nets have formed our pathway to topologies. With some justification, it might be said that in our setup topologies form a tool to study convergence of nets. It is only fair to point out alternatives. This we do in 18.2 and 18.3.

Next, we have striven to avoid a surfeit of terminology. The vocabulary of even elementary parts of Topology is staggering and its use in the literature is annoyingly inconsistent. We have attempted to stick to the bare necessities. If you are going to study the subject in some depth, you will need many technical terms we have not mentioned. In 18.4 through 18.8, we present an abbreviated dictionary.

18.2

A frequent alternative to nets is formed by filters. A *filter* on a set X is a collection λ of subsets of X with the following properties:

$$\text{If } A \in \lambda \text{ and } A \subset B \subset X, \text{ then } B \in \lambda;$$

$$\text{If } A_1, A_2 \in \lambda, \text{ then } A_1 \cap A_2 \in \lambda;$$

$$\varnothing \notin \lambda.$$

If, on X, a topology ω is given, a filter λ is said to *converge* to a point a of X if every neighborhood of a belongs to λ. The converging filters determine the topology. Indeed, a subset U of X is open if and only if U belongs to every filter that converges to a point of U.

18.3

Yet another approach to Topology is due to Casimir Kuratowski in his book *Topology*. For a set X, we assume that there is given a so-called "closure operator" which assigns to each subset Y of X another subset, denoted Y^Δ, such that the following rules are obeyed:

$$\varnothing^\Delta = \varnothing;$$

$$Y \subset Y^\Delta \text{ for every } Y;$$

$$Y^{\Delta\Delta} = Y^\Delta \text{ for every } Y;$$

$$(Y \cup Z)^\Delta = Y^\Delta \cup Z^\Delta \text{ for every } Y \text{ and } Z.$$

Then the collection

$$\omega_\Delta := \{X \backslash Y \ : \ Y = Y^\Delta\}$$

is a topology on X, and for every Y, Y^Δ is the ω_Δ-closure of Y. Conversely, if ω is any topology on X, then the operator $Y \mapsto Y^-$ is a closure operator in the above sense.

An interesting puzzle associated with taking complements and closures is due to Kuratowski also. He showed that by starting with one subset of a topological space and repeatedly applying closure and complement operations, one can get at most 14 sets. The puzzle is: Find subsets of $\mathbb{R}$ for which exactly 14 sets result.

In the following, X is a topological space.

Countability Conditions

18.4

Let $a \in X$. We say that X has a *countable base at a* if there exists a sequence $V_1, V_2, \ldots$ of neighborhoods of a such that every neighborhood of a contains a V_n.

X is said to be *first countable* (or to *satisfy the first axiom of countability*) if it has a countable base at each of its points.

X is *second countable* (or *satisfies the second axiom of countability*) if its topology has a countable base.

X is *separable* if it contains a countable dense subset (e.g., $\mathbb{R}$ is separable, as $\mathbb{Q}$ is countable and dense in $\mathbb{R}$).

Every second countable space is first countable; so is every metrizable space. For metrizable spaces, the second axiom of countability is equivalent to separability. See Exercise 18.A.

Separation Conditions

18.5

Besides Hausdorffness, many other "separation" properties have been singled out. Unfortunately, there is little agreement as to the terminology. We adopt the formulations of Kelley's *General Topology*, but we warn the reader that there is no uniformity in the literature.

X is called *regular* if for every closed subset A of X and for every $b \in X \setminus A$, there exist disjoint open sets U and V such that $A \subset U$ and $b \in V$.

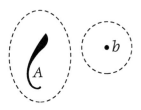

X is *completely regular* (see Exercise 14.B) if for every closed subset A of X and every $b \in X \setminus A$, there exists a continuous function $f : X \to \mathbb{R}$ such that $f = 0$ everywhere on A but $f(b)$ is 1.

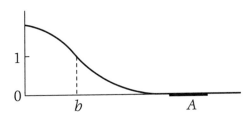

Every completely regular space is regular. (Take $U := \{x : f(x) < \frac{1}{2}\}$, $V := \{x : f(x) > \frac{1}{2}\}$.) Every metrizable space is completely regular. See Exercise 18.B.

X is *normal* (15.4) if for every pair of disjoint closed subsets A and B, there exist disjoint open sets U and V such that $A \subset U$ and $B \subset V$.

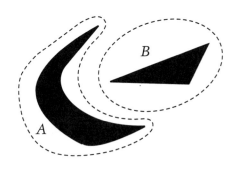

Metrizable spaces and compact Hausdorff spaces are normal (Theorem 15.7).

X is a T_1-*space* if for all $a, b \in X$ with $a \neq b$ there exists an open set U with $a \in U$ and $b \notin U$. This is the case if and only if all singleton sets $\{c\}$ are closed. Hausdorff spaces (sometimes called T_2-*spaces*) are T_1. Normal T_1-spaces are completely regular [Urysohn's Lemma (15.6)].

There are also T_0-spaces, $T_{3\frac{1}{2}}$-spaces, and an entire zoo of $T_{\text{something}}$-spaces, not to mention Fréchet spaces, Urysohn spaces, completely normal spaces, fully normal spaces, and perfectly normal spaces, but we are not going to bother with them.

Compactness Conditions

18.6

X is *sequentially compact* (13.1) if every sequence in X has a convergent subsequence. Some sequentially compact spaces are not compact (Example 13.5); some compact spaces are not sequentially compact (Example 17.13).

On a compact space, every continuous function is bounded [Theorem 13.16(iii)]. In general, X is called *pseudocompact* if all continuous functions on X are bounded. Every sequentially compact space is pseudocompact. (Copy the proof of Theorem 3.9.) Hence, there exist noncompact pseudocompact spaces.

X is *countably compact* if every cover of X by countably many open sets has a finite subcover. X is *Lindelöf* if every cover of X by open sets has a countable subcover. Thus,

$$\text{compact} = \text{countably compact} + \text{Lindelöf}.$$

The following two definitions have a slightly different character. X is *locally compact* if every point of X has a compact neighborhood. (Prime examples are, of course, $\mathbb{R}$ and $\mathbb{R}^2$.) X is *σ-compact* if it is a union of countably many compact subsets. (The prefix "σ-" is often used to indicate countability.)

Compactifications

18.7

A compact Hausdorff space $\overline{X}$ is called a *compactification* of X if X is a dense subset of $\overline{X}$.

Only completely regular Hausdorff spaces have compactifications [(i) and (iii) of Exercise 18.B].

If X is a compact Hausdorff space, its only compactification is X itself. A locally compact Hausdorff space X that is not compact has a compactification obtainable by adding one point to X; see Exercise 18.G.

At the other end of the spectrum, there is a compactification that is, in some way, as large as possible. Suppose X is completely regular. Then, there exists an essentially unique compactification βX of X, its *Stone-Čech compactification*, with the property that every continuous map of X into any compact Hausdorff space Y can be extended to a continuous map $\beta X \rightarrow Y$. (In particular, if $\overline{X}$ is any compactification of X, then the identity map $X \rightarrow \overline{X}$ extends to a continuous map $\beta X \rightarrow \overline{X}$, which is easily seen to be surjective. In this sense, βX is the largest compactification of X.) Even for a simple space like $\mathbb{N}$, the Stone-Čech compactification is an extremely complex object. For more information, we advise the reader to turn to the glorious book *Rings of Continuous Functions* by L. Gillman and M. Jerison and its sequel, *The Stone-Čech Compactification* by J. Walker.

Connectivity Conditions

18.8
In Chapter 16, we have already considered connectedness (16.1) and path connectedness (16.5).

We say that X is *locally (path) connected* if the (path) connected open subsets form a base for the topology.

X it *totally disconnected* if no component of X (see 15.12) contains more than one element; it is *zero-dimensional* if a base for the topology is formed by the subsets that are both open and closed. Hausdorff zero-dimensional spaces are totally disconnected.

18.9
Many other topological terms are to be found in the standard texts such as the ones by Kelley and Dugundji. If you are interested in Patho-topology (Does there exist a locally compact and pseudo-compact zero-dimensional Hausdorff space that is not compact?), you will enjoy *Counterexamples in Topology* by L.A. Steen and J.A. Seebach. This wonderful little book presents several diagrams depicting valid implications and a host of concrete topological spaces with unexpected properties. (Example 15.H is one of those.)

Extra: Dates from the History of General Topology

1619 Descartes discovers what is now called the Euler-Poincaré formula. (See the Extra of Chapter 16.)

1676 Gottfried Wilhelm Leibniz (1646-1716) uses the words *geometria situs* (geometry of location) and foresees the importance of its study.

1687 Newton's *Principia Mathematica* on calculus.

1737 Leonhard Euler (1707-1783) discusses the problem of the Koenigsberg bridges.

1799 Carl Friedrich Gauss (1781-1848) publishes his thesis containing the first proof of the Main Theorem of Algebra, making a passing remark on topological aspects.

1805 Bernhard Bolzano publishes his thesis on geometry and starts modern rigor in analysis. (Intermediate Value Theorem: $\mathbb{R}$ is connected.)

1833 Gauss states: Of the geometry of position which Leibniz had initiated and to which it remained for only two geometers, Euler and Vandermonde, to throw a feeble glance, we know and possess, after a century and a half, very little more than nothing.

1852 The Four-Color Problem is formulated.

1865 August Ferdinand Moebius (1790-1868) studies the Moebius strip.

1866 Death of Bernhard Riemann, the first to use topological techniques in analysis.

1873 James Clerk Maxwell uses topology in his study of electromagnetic fields.

1878 Georg Cantor (1845-1918) proves that for all n and m there exists a bijection of $\mathbb{R}^n$ onto $\mathbb{R}^m$.

1895 First paper fully devoted to the theory of Topology by Henri Poincaré (1845-1912) with the title *Analysis Situs* and introducing algebraic methods.

1906 Maurice Fréchet (1878.1973) in his thesis introduces abstract metric spaces.

1911 Luitzen Brouwer proves that $\mathbb{R}^n$ and $\mathbb{R}^m$ are homeomorphic only if $n = m$.

1914 Hausdorff's book "Mengenlehre."

1920 Stefan Banach introduces normed vector spaces.

1923 Alexandrov and Urysohn introduce compactness.

1937 Čech proves the Tychonoff Theorem.

1976 Appel and Haken prove the Four-Color Theorem.

The following are the main architects of the theory expounded in this book. With each name we mention one or two keywords, indicating a connection with Topology, not necessarily the most important contribution to Mathematics.

Paul Alexandroff: 1896-1983, Russia, Compactness, contributions to Algebraic Topology.

Kenneth Appel: 1932- , United States, Four-Color Theorem.

René Baire: 1874-1932, France, Baire Category Theorem.

Stefan Banach: 1882-1945, Poland, Banach's Contraction Principle.

Bernhard Bolzano: 1781-1848, Prague (now in the Czech Republic), Bolzano-Weierstrass Theorem.

Emile Borel: 1871-1956, France, Heine-Borel Theorem.

Luitzen Brouwer: 1882-1966, the Netherlands, Brouwer Fixed Point Theorem, Intuitionism.

Georg Cantor: 1845-1918, Germany (Halle), Cantor set, founder of Set Theory.

Augustin-Louis Cauchy, 1789-1857, France, Foundations of Analysis.

Paul Cohen: 1934- , United States, Continuum Hypothesis, Axiom of Choice.

Richard Dedekind: 1831-1916, Switzerland, Germany (Braunschweig), Dedekind's Axiom.

Peter Lejeune Dirichlet: 1805-1859, Poland, Germany (Berlin, Goettingen), Definition of "function," Analysis.

Leonhard Euler: 1707-1783, Russia (St. Petersburg), Germany (Berlin), Koenigsberg Bridges, Analysis.

Abraham Fraenkel: 1891-1965, Germany, Israel, Set Theory.

Carl Friedrich Gauss: 1777-1855, Germany (Goettingen), Fundamental Theorem of Algebra.

Hans Hahn: 1879-1934, Austria, Germany, Extension theorems for continuous functions.

Wolfgang Haken: 1928- , United States, Four-Color Theorem.

Felix Hausdorff: 1868-1942, Germany, Axioms for Hausdoff topological spaces.

Eduard Heine: 1821-1881, Germany (Halle), Heine-Borel Theorem, Uniform Continuity.

David Hilbert: 1862-1943, Germany (Goettingen), Hilbert Cube.

Camille Jordan: 1838-1922, France, Closed Curve Theorem.

Felix Klein: 1849-1925, Germany (Erlangen, Goettingen), Klein Bottle.

Rudolph Lipschitz: 1832-1903, Germany (Berlin, Bonn), Lipschitz condition.

John Listing: 1808-1882, Germany (Goettingen), First book on Topology.

James Clerk Maxwell: 1831-1879, United Kingdom (Aberdeen, London), Vector fields, founder of Electromagnetism.

August Moebius: 1790-1868, Germany, Moebius Band.

Henri Poincaré: 1854-1912, France, Founder of Algebraic Topology.

Bernhard Riemann: 1826-1866, Germany (Goettingen), Surfaces.

Hermann Schwarz: 1843-1921, Switzerland, Germany (Goettingen, Berlin), Schwarz Inequality.

Karl Weierstrass: 1815-1897, Germany (Berlin), Exact Analysis.

Ernst Zermelo: 1871-1953, Germany (Goettingen), Switzerland, Set Theory.

Max Zorn: 1906-1993, United States, Zorn's Lemma.

Exercises

18.A. (Concerning separability and the second axiom of countability; see Theorem 15.16.)
  (i) Show that *the second countability axiom implies separability.*
  (ii) Show that *every totally bounded metric space is separable.* (In particular, *compact metrizable spaces are separable.*)
  (iii) Show that *if (X,d) is a metric space and D is a dense subset of X, then the balls $B_{1/n}(a)$ with $n \in \mathbb{N}$, $a \in D$ form a base for the d-topology.*
  (iv) Use (i) and (iii) to prove that *a metrizable space is separable if and only if it is second countable.*
  (v) Deduce that *if X is a metrizable space, then every subset of X is separable under the relative topology.*
  (vi) For nonmetrizable spaces, one has to be more careful. Consider the following collection ω of subsets of $\mathbb{R}$:

  $$A \in \omega \iff 0 \in A \text{ or } A = \varnothing.$$

  Show that ω is a topology on $\mathbb{R}$. Show that $\{0\}$ is ω-dense in $\mathbb{R}$, so that $(\mathbb{R}, \omega)$ is a separable space, but that $\mathbb{R} \backslash \{0\}$ is not separable under the relative topology.

18.B. (Completely regular spaces; see also Exercise 14.B.)
  (i) Show that *compact Hausdorff spaces are completely regular* (Urysohn's Lemma).
  (ii) Show, without using deep theory such as Urysohn's Lemma, that *metrizable spaces are completely regular.*
  (iii) Show that *subspaces of completely regular spaces are completely regular.*
  (iv) Let S be a set. Show that $\mathbb{R}^S$ *is completely regular under the product topology.*

18.C. (Countably compact spaces.)
(i) Show that X is *countably compact if and only if the following is true*:

$$\text{If } A_1, A_2, \ldots \text{ are nonempty closed sets}$$
$$\text{and } A_1 \supset A_2 \supset \ldots, \text{ then } \bigcap_n A_n \neq \varnothing.$$

(ii) Deduce that *sequentially compact spaces are countably compact.*
(iii) Show that *countably compact spaces are pseudocompact.*
(iv) Show that *for metrizable spaces, compactness, sequential compactness, pseudocompactness and countable compactness are the same.* [Hint: Show, if $a_1, a_2, \ldots$ is a sequence with no converging subsequence, then $\{a_n, a_{n+1}, a_{n+2}, \ldots\}$ is closed for every n. Deduce that countable compactness implies (sequential) compactness. Now, assume X is not countably compact. By (i), there exist nonempty closed sets $A_1 \supset A_2 \supset \cdots$ with empty intersection. For each n, choose a continuous $f_n : X \to [0, 2^{-n}]$ with $A_n = \{x : f_n(x) = 0\}$ [Exercises 5.K and 6.E].) Then, $1/\sum f_n$ is an unbounded continuous function.]

18.D. (Local compactness, σ-compactness.) Show that $\mathbb{Q}$ is σ-compact, not locally compact.

18.E. (The Lindelöf property and the second countability axiom.)
(i) Show that *second countable spaces are Lindelöf.* (Let $\{U_n : n \in \mathbb{N}\}$ be a countable base for the topology of X and let φ be a cover by open sets. Let $N := \{n \in \mathbb{N} : U_n$ is contained in a set that belongs to $\varphi\}$. For $n \in N$, choose $W_n \in \varphi$ with $U_n \subset W_n$. Show that $\{W_n : n \in N\}$ is a cover of X.)
(ii) Show that *a Lindelöf metric space is separable.* (For $p \in \mathbb{N}$, cover X by balls with radius p^{-1}.)
(iii) Now use Exercise 18.A(iv) to prove that *for metrizable spaces the notions of Lindelöf, second countable, and separable are equivalent.*

18.F. (Locally compact Hausdorff spaces.)
(i) Prove that *if X is a locally compact Hausdorff space and $a \in X$, then every neighborhood of a contains a compact neighborhood of a.* [Hint: Let K be a compact neighborhood of a, let $U \subset X$ be open, and $a \in U$; we wish to show that U contains a compact neighborhood K_0 of a. Show that we may assume $U \subset K$. Apply Urysohn's Lemma to obtain a continuous function f on K with $f(a) = 1$, $f = 0$ everywhere on $K\backslash U$. Show that $K_0 = \{x \in K : f(x) \geq \frac{1}{2}\}$ satisfies the requirements.]
(ii) Use (i) to prove that *every open subset of a locally compact Hausdorff space is locally compact (in the relative topology).*

18.G. (The one-point compactification.)
Let (X, ω) be a noncompact locally compact Hausdorff space, e.g., $\mathbb{R}^2$. Let ∞ be a mathematical object that is not an element of X and put $\alpha X := X \cup \{\infty\}$. Show that

$$\overline{\omega} := \omega \cup \{\alpha X \backslash K : K \text{ is a compact subset of } X\}$$

is a topology on αX, that the topological space $(\alpha X, \overline{\omega})$ is compact Hausdorff, that X is dense in αX, and that ω is the relative topology on X induced by $\overline{\omega}$.

The space $(\alpha X, \overline{\omega})$ is called the *one-point* (or *Alexandroff*) *compactification* of (X, ω). It is not hard to see that the one-point compactification of $\mathbb{R}$ is homeomorphic to a circle, that of $\mathbb{R}^2$ to the sphere $\{x \in \mathbb{R}^3 : \|x\| = 1\}$.

18.H. (Local connectedness.)

   (i) Observe that a discrete space with at least two points is locally connected but not connected.

   (ii) In Example 16.7, we saw that the "topologists comb without zero" is connected. Prove that it is not locally connected.

18.I. (Disconnectedness.)

   (i) Show that $\mathbb{Q}$ *and the space considered in Examples 5.3(v) and 5.5(v) are zero-dimensional.*

   (ii) Show that *a compact Hausdorff space is zero-dimensional if and only if it is totally disconnected.* (Use Exercise 16.E.)

   (iii) Show that *every open subset of a locally connected space is locally connected.*

   (iv) *Let X be a locally connected space. Prove that every connected component of X is open.*

19

CHAPTER

Countable Sets

19.1

If X is the set of people I want to invite for dinner and Y is the set of my chairs, then it is important for me to know whether X and Y have the same size. The obvious way to find out is to count them.

But what if X and Y are infinite sets? Does it make sense, for example, to ask whether there are as many positive integers as they are real numbers in the interval $[0, 1]$? At first sight, it might seem that nothing can be said except that both $\mathbb{N}$ and $[0, 1]$ have infinitely many elements.

19.2

Georg Cantor took a different view. Even for finite sets, counting is a brute-force method that is not always the natural one. Suppose X and Y consist of the left and the right legs of a millipede,

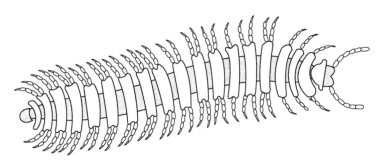

or X is the set of all prime numbers less than 10^9 and Y is the set of their squares. Then, we see at a glance that X and Y are equally numerous, simply by pairing each element of X with one element of Y. Mathematically speaking, we make a bijective map of X onto Y:

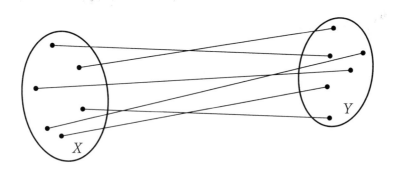

Extending this idea to infinite sets, Cantor proposed to call two sets *equipollent* if there exists a bijective map between them and to interpret equipollency of sets as their "having the same size."

It then becomes a nontrivial question whether $\mathbb{N}$ and $[0, 1]$ "have the same size" and, interestingly, it turns out that they do *not*: $[0, 1]$ is strictly "larger" than $\mathbb{N}$. [See Example 19.5(iii).] Thus, Cantor's definition enables us to distinguish various kinds of infinity!

19.3
It leads to unexpected phenomena.

Let E be the set of all squares of positive integers:

$$E := \{1, 4, 9, 16, \ldots\}.$$

Centuries before Cantor, Galileo had already pointed out that although E seems much smaller than $\mathbb{N}$, there exists a bijection between them:

$$
\begin{array}{ccccc}
1 & 2 & 3 & 4 & \ldots \\
\updownarrow & \updownarrow & \updownarrow & \updownarrow & \\
1 & 4 & 9 & 16 & \ldots
\end{array}
$$

In a similar way, the interval $(0, 1)$ is only a small portion of $(0, \infty)$, but the formula

$$x \mapsto \frac{x}{1 - x} \qquad (0 < x < 1)$$

establishes a bijection between them: $(0, 1)$ and $(0, \infty)$ are "equipollent." The following picture illustrates the situation:

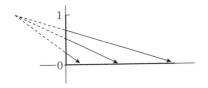 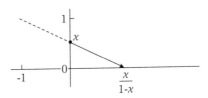

In this sense, "equipollency" does not reflect our intuitive idea of "equal size." Nevertheless, the mathematical world has, after a considerable struggle, accepted Cantor's suggestion. Equipollency has become an important concept.

This section of the book considers in particular sets that are equipollent with the set $\mathbb{N}$ of all positive integers.

19.4

An "enumeration" of a set X is an arrangement of the elements of X into a finite or infinite sequence

$$x_1, x_2, \ldots$$

without repetitions. More formally, the definition runs as follows.

First we define the *initial intervals of* $\mathbb{N}$ to be

- $\mathbb{N}$ itself,
- the empty set, and
- the sets $\{1, 2, \ldots, n\}$ where $n \in \mathbb{N}$.

An *enumeration* of a set X is a bijective map of some initial interval of $\mathbb{N}$ onto X. Such a bijection f effectively puts the elements of X into a sequence without repetitions:

$$f(1), f(2), \ldots .$$

A set X is called *countable* (or *enumerable*) if there exists such an enumeration, i.e., if X is equipollent with some initial interval of $\mathbb{N}$. Other sets are *uncountable*. We may distinguish two types of countable sets:

- the finite sets (e.g., the empty set);
- the sets that are equipollent with $\mathbb{N}$ itself. These are said to be *countably infinite*.

If X and Y are equipollent sets and one of them is countable, then so is the other. Every pair of countably infinite sets is equipollent.

Examples 19.5

(i) Trivially, $\mathbb{N}$ is countably infinite. By Galileo's paradox, the set of all squares of positive integers is countably infinite. So is the set $\mathbb{Z}$ of all integers, as one sees from the formula

$$\mathbb{Z} = \{0, 1, -1, 2, -2, 3, -3, \ldots\}.$$

More formally, one can define $f : \mathbb{N} \to \mathbb{Z}$ by

$$f(n) := \frac{n}{2} \quad \text{if } n \text{ is even,}$$

$$f(n) := \frac{1-n}{2} \quad \text{if } n \text{ is odd,}$$

and show that f is a bijection.

(ii) Finite sets are not countably infinite, of course. More interesting is the observation (also due to Cantor) that *the collection $\mathcal{P}(\mathbb{N})$ of all subsets of $\mathbb{N}$ is not countable*. The proof of this statement is so simple that it leaves one with a feeling of being cheated: *Suppose* there is a bijection f of $\mathbb{N}$ onto $\mathcal{P}(\mathbb{N})$. Consider the set

$$A := \{n \in \mathbb{N} : n \notin f(n)\}.$$

Take any positive integer n. If $n \notin f(n)$, then $n \in A$, so $A \neq f(n)$; if $n \in f(n)$, then $n \notin A$, and again $A \neq f(n)$. Thus, for every positive integer n, we obtain $A \neq f(n)$. But then, f is not surjective. *Contradiction!*

Observe that the injectivity of f has not played a role. We have actually proved that no map $\mathbb{N} \to \mathcal{P}(\mathbb{N})$ can be surjective.

(iii) In a similar way, we show that *the interval $[0,1]$ is uncountable* by proving that there is no surjection of $\mathbb{N}$ onto $[0, 1]$. Let $f : \mathbb{N} \to [0, 1]$; we will obtain an element of $[0, 1]$ that is no value of f.

First, choose a subinterval $[a_1, b_1]$ of $[0, 1]$ with

$$b_1 - a_1 \leq 5^{-1}, \qquad f(1) \notin [a_1, b_1].$$

If you see that this can be done, you will agree that there is a subinterval $[a_2, b_2]$ of $[a_1, b_1]$ with

$$b_2 - a_2 \leq 5^{-2}, \qquad f(2) \notin [a_2, b_2],$$

and that we can continue and form a sequence of intervals

$$[0, 1] \supset [a_1, b_1] \supset [a_2, b_2] \supset \cdots$$

such that for all n

$$b_n - a_n \leq 5^{-n}, \qquad f(n) \notin [a_n, b_n].$$

By Cantor's Theorem, (2.19), there is a number c that lies in every $[a_n, b_n]$. Then, $c \in [0, 1]$ and $f(n) \neq c$ for all n, so that f cannot be surjective.

(iv) *Every subset of $\mathbb{N}$ is countable.* Proof: Let $X \subset \mathbb{N}$. As every finite set is countable, we may assume X is not finite. Then, for every finite subset Y of X the set $X\backslash Y$ is a nonempty subset of $\mathbb{N}$, hence has a smallest element (Exercise 2.F). Now define $f : \mathbb{N} \to \mathbb{N}$ as follows:

$f(1) :=$ the smallest element of X,

$f(2) :=$ the smallest element of $X\backslash\{f(1)\}$,

$f(3) :=$ the smallest element of $X\backslash\{f(1), f(2)\}$,

etc.

Then, f is an injection $\mathbb{N} \to X$; we are done if f is surjective. *Suppose* $p \in X$ and $p \notin \{f(1), f(2), \ldots\}$. Then for every $n \in \mathbb{N}$, we have

$$p \in X\backslash\{f(1), f(2), \ldots, f(n-1)\}$$

and therefore $f(n) \leq p$. Then, all values of f lie in the finite set $\{1, 2, \ldots, p\}$. *Contradiction.*

Theorem 19.6
Let X and Y be sets; let X be countable. Then in each of the following situations Y is countable:

(i) *There exists an injective map $Y \to X$.*
(ii) *There exists a surjective map $X \to Y$.*

[It follows from (i) that every subset of a countable set is countable.]

Proof
Choose a bijection $f : I \to X$ where I is a suitable initial interval of $\mathbb{N}$.

(i) Let $g : Y \to X$ be injective.
   Then $f^{-1} \circ g$ is an injection $Y \to \mathbb{N}$, hence a bijection $Y \to A$ where

$$A := \{(f^{-1} \circ g)(y) : y \in Y\}.$$

   Y and A are equipollent and A is countable [Example 19.5(iv)]; then so is Y.
(ii) Let $h : X \to Y$ be surjective.

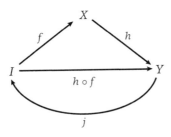

$h \circ f$ is a surjection $I \to Y$. Define $j : Y \to I$ by

$$j(y) := \text{ the smallest element of } \{n \in I : (h \circ j)(n) = y\}.$$

Then j is an injection of Y into the countable set I, so Y is countable by (i) of this theorem. ∎

Example 19.7

$\mathbb{N} \times \mathbb{N}$, *the set of all pairs (m,n) with $m,n \in \mathbb{N}$, is countable.* Indeed, the formula

$$(m, n) \longmapsto 2^m 3^n$$

establishes an injection $\mathbb{N} \times \mathbb{N} \to \mathbb{N}$.

The following diagram suggests an enumeration of $\mathbb{N} \times \mathbb{N}$ (that has nothing to do with the injection we just mentioned):

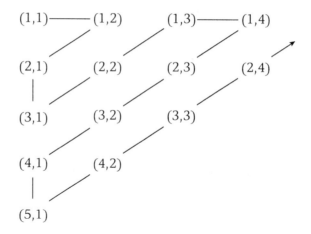

Theorem 19.8

If each of the sets $X_1, X_2, \ldots$ is countable, then so is their union.

Proof

For each n, choose an initial interval I_n of $\mathbb{N}$ and a bijection $f_n : I_n \to X_n$. Define $J \subset \mathbb{N} \times \mathbb{N}$ by

$$J := \{(n, i) : i \in I_n\}$$

and define $g : J \to \bigcup_{n \in \mathbb{N}} X_n$:

$$g(n, i) := f_n(i) \qquad (n \in \mathbb{N}; \ i \in I_n).$$

As a subset of the countable set $\mathbb{N} \times \mathbb{N}$, J is countable. Furthermore, g is surjective. Hence, $\bigcup_{n \in \mathbb{N}} X_n$ is countable according to Theorem 19.6(ii).

(Note that g might not be *injective*, as the sets $X_1, X_2, \ldots$ might not be pairwise disjoint.) ∎

Example 19.9
$\mathbb{Q}$ *is countable*. Indeed, for $n \in \mathbb{N}$ let

$$Q_n := \{ \tfrac{k}{n} : k \in \mathbb{Z} \}.$$

Each Q_n is countable, being equipollent with $\mathbb{Z}$ [see Example 19.5(i)] and $\mathbb{Q} = \bigcup_{n \in \mathbb{N}} Q_n$.

Extra: The Continuum Hypothesis

We have seen that $[0, 1]$ is not countable. It follows that $[0, 1]$ is not a subset of any countable set, so that, e.g., $\mathbb{R}$ cannot be countable. Actually, $\mathbb{R}$ and $[0, 1]$ are equipollent. We can see that as follows.
   For two sets, A and B, let us write

$$A \equiv B$$

if A and B are equipollent. If A, B, and C are sets, $A \equiv B$, and $B \equiv C$, then $A \equiv C$.
   It is not difficult to make a bijection (or even a homeomorphism) between $\mathbb{R}$ and the interval $(0, 1)$:

$$\mathbb{R} \equiv (0, 1).$$

Consider the function $f : \mathbb{R} \to \mathbb{R}$ defined by

$$f(\tfrac{1}{n}) := \tfrac{1}{n+1} \qquad (n = 1, 2, 3, \ldots),$$
$$f(x) := x \quad \text{if} \quad x \notin \{1, \tfrac{1}{2}, \tfrac{1}{3}, \ldots\}.$$

f is injective, so $f(X) \equiv X$ for every $X \subset \mathbb{R}$. Now, f maps $[0, 1]$ onto $[0, 1)$ and $(0, 1]$ onto $(0, 1)$. As, obviously, $[0, 1) \equiv (0, 1]$, we obtain

$$\mathbb{R} \equiv (0, 1) \equiv (0, 1] \equiv [0, 1) \equiv [0, 1].$$

Less simple, but still provable is:

$$\mathbb{R} \equiv \mathcal{P}(\mathbb{N}).$$

If A and B are sets, we put

$$A \leq B$$

if A is equipollent with a subset of B. Clearly, this is the case if and only if there exists an injective map of A into B. Therefore,

$$A \leq B, \; B \leq C \implies A \leq C.$$

A set A is infinite if and only if $\mathbb{N} \leq A$.

A famous result in Set Theory is the Cantor-Schroeder-Bernstein Theorem:

$$A \leq B, \ B \leq A \implies A \equiv B.$$

We outline a proof:

As A is equipollent with a subset of B, we may as well assume that A is a subset of B. Let g be an injective map $B \to A$. Define

$$C := B \backslash A;$$

$$X := C \cup g(C) \cup g(g(C)) \cup g(g(g(C))) \cup \ldots;$$

$$Y := B \backslash X.$$

Then $X = C \cup g(X)$. The sets C and $g(X)$ are disjoint, because $g(X) \subset g(B) \subset A$. Thus,

$$C, g(X) \text{ and } Y \text{ are pairwise disjoint.}$$

Now, $g(X) \equiv X$. As $g(X) \cap Y = \varnothing$ and $X \cap Y = \varnothing$, it follows that $g(X) \cup Y \equiv X \cup Y$. But, $g(X) \cup Y = (X \backslash C) \cup (B \backslash X) = B \backslash C = A$, so $A \equiv B$.

If X is an infinite subset of $\mathbb{R}$, then $\mathbb{N} \leq X \leq \mathbb{R}$. The *Continuum Hypothesis* (**CH**) is the statement:

Every infinite subset of $\mathbb{R}$ is equipollent with either $\mathbb{N}$ or $\mathbb{R}$.

In other words: Every uncountable subset of $\mathbb{R}$ is equipollent with $\mathbb{R}$ itself.

For a long time, one of the celebrated problems of Set Theory was: Is the Continuum Hypothesis true? In a famous lecture in 1900, David Hilbert published a list of the most important problems of the mathematics of that time. (Most have been solved since then.) His first question was about the truth of the Continuum Hypothesis. The answer turns out to be yes and no. Although **CH** might seem different in character from the Axiom of Choice we discussed in Chapter 17, the situation is the same: In 1938, Kurt Gödel showed that **CH** is consistent with **ZF** and even with **ZF + AC**; in 1963 Paul Cohen proved the same for the negation of **CH**. In **ZF**, the Continuum Hypothesis is both irrefutable and unprovable!

The Continuum Hypothesis has curious consequences. We mention one. To get its import you should realize that countable sets are really very small in comparison with $\mathbb{R}$. Indeed, if X is a countable subset of $\mathbb{R}$, then $X \backslash \mathbb{R}$ is always equipollent with $\mathbb{R}$ itself. (If you own as many dollars as there are real numbers, you may lose an infinite amount of money and you will never know the difference.)

Sierpiński has shown that the Continuum Hypothesis entails the existence of a subset A of $\mathbb{R}^2$ such that

for every horizontal line L the set $L \cap A$ is countable,

for every vertical line M the set $M \backslash A$ is countable.

In other words, you can paint part of the plane red and the rest blue in such a way that every horizontal line is almost entirely red and every vertical line is almost entirely blue!

Further Reading

Smullyan, R., *Satan, Cantor and Infinity*, Alfred A. Knopf, New York 1992.

A Farewell to the Reader

There are two classics that every budding topologist or analyst should take in hand from time to time. These are the books by Kelley and Dugundji listed in the "Literature." In fact, these are the works that professional mathematicians refer to, when needed, in their research papers. Kelley's *General Topology* is intended for an education with Analysis in mind primarily, while Dugundji's book also contains a treatment of Algebraic Topology.

For examples and counterexamples of a rich variety of properties, there is the book by Steen and Seebach: *Counterexamples in Topology*.

Gillman and Jerison's *Rings of Continuous Functions* pursues in grandeur the interactions between topological properties of a space and algebraic properties of the set of all continuous functions on it. Highly recommended reading!

If you have become interested in Set Theory, there is no better place to start reading than P. Halmos' *Naive Set Theory*.

A subject we have occasionally come across is Algebraic Topology. If you want to know more about it, consider the books by Rotman and Greenberg and Harper mentioned in the Extra of Chapter 16.

For the historical development of mathematics, one should consult the books by Bell and Kline but not without a warning: Bell's writing is wonderful but not always reliable in its facts. The history of Topology does not seem well researched. For a starting point to more information the reader might want to consult the articles *Topology: Geometric, Algebraic* by E. Scholz (pages 927-938 in *Companion Encyclopedia of the History of Philosophy of the Mathematical Sciences*, Volume 2, Routledge Inc., London

1994) and *Topology: Invariance of Dimension* by Joseph W. Dauben (pages 939-946 in the same Encyclopedia).

Once you have come as far as the end of this book, assuming you have enjoyed the journey, you will enjoy every new (and old) issue of the *American Mathematical Monthly* and the *Mathematical Intelligencer*. The authors of this book consider these their favorite journals. As to the present book, they were particularly inspired by recent proofs of the Jordan Closed Curve Theorem [Monthly 91, (1984), 641-643], Zorn's Lemma from the Axiom of Choice [Monthly 98, (1991), 353-354], Tychonoff's Theorem [Monthly 99, (1992), 932-934] and an example of a nowhere differentiable continuous function [Monthly 98, (1991), 411-416].

Literature

1. Dugundji, J., *Topology*, Allyn and Bacon, Boston, 1966.
2. Gillman, L., and M. Jerison, *Rings of Continuous Functions*, Springer-Verlag, New York, 1976.
3. Halmos, P., *Naive Set Theory*, Springer-Verlag, New York, 1974.
4. Hausdorff, F., *Set Theory*, Chelsea Publishing Company, New York, 1957.
5. Jech, T.J., *The Axiom of Choice*, North-Holland Publishing Company, Amsterdam, 1973.
6. Kelley, J.L., *General Topology*, Springer-Verlag, New York, 1975.
7. Kuratowski, C., *Topology I, II*, Academic Press, New York - London, 1955.
8. Steen, L.A., and J.A. Seebach, Jr., *Counterexamples in Topology*, Springer-Verlag, New York, 1978.

Index of Symbols

Index of Terms

Undergraduate Texts in Mathematics

(continued from page ii)